Decommissioned Submarines in the Russian Northwest

NATO ASI Series

Advanced Science Institutes Series

A Series presenting the results of activities sponsored by the NATO Science Committee, which aims at the dissemination of advanced scientific and technological knowledge, with a view to strengthening links between scientific communities.

The Series is published by an international board of publishers in conjunction with the NATO Scientific Affairs Division

A Life Sciences	Plenum Publishing Corporation
B Physics	London and New York
C Mathematical and Physical Sciences	Kluwer Academic Publishers
D Behavioural and Social Sciences	Dordrecht, Boston and London
E Applied Sciences	
F Computer and Systems Sciences	Springer-Verlag
G Ecological Sciences	Berlin, Heidelberg, New York, London,
H Cell Biology	Paris and Tokyo
I Global Environmental Change	

PARTNERSHIP SUB-SERIES

1. Disarmament Technologies	Kluwer Academic Publishers
2. Environment	Springer-Verlag / Kluwer Academic Publishers
3. High Technology	Kluwer Academic Publishers
4. Science and Technology Policy	Kluwer Academic Publishers
5. Computer Networking	Kluwer Academic Publishers

The Partnership Sub-Series incorporates activities undertaken in collaboration with NATO's Cooperation Partners, the countries of the CIS and Central and Eastern Europe, in Priority Areas of concern to those countries.

NATO-PCO-DATA BASE

The electronic index to the NATO ASI Series provides full bibliographical references (with keywords and/or abstracts) to more than 50000 contributions from international scientists published in all sections of the NATO ASI Series.
Access to the NATO-PCO-DATA BASE is possible in two ways:

– via online FILE 128 (NATO-PCO-DATA BASE) hosted by ESRIN,
Via Galileo Galilei, I-00044 Frascati, Italy.

– via CD-ROM "NATO-PCO-DATA BASE" with user-friendly retrieval software in English, French and German (© WTV GmbH and DATAWARE Technologies Inc. 1989).

The CD-ROM can be ordered through any member of the Board of Publishers or through NATO-PCO, Overijse, Belgium.

Series 2: Environment – Vol. 32

Decommissioned Submarines in the Russian Northwest

Assessing and Eliminating Risks

edited by

Elizabeth J. Kirk

American Association for the Advancement of Science,
Washington, D.C., U.S.A.

Springer-Science+Business Media, B.V.

Proceedings of the NATO Advanced Research Workshop on
Recycling, Remediation and Restoration Strategies for
Contaminated Civilian and Military Sites in the Arctic Far North
Kirkenes, Norway
24–28 June 1996

A C.I.P. Catalogue record for this book is available from the Library of Congress.

ISBN 978-94-010-6368-5   ISBN 978-94-011-5618-9 (eBook)
DOI 10.1007/978-94-011-5618-9

Printed on acid-free paper

TABLE OF CONTENTS

ACKNOWLEDGMENTS

This book resulted from a NATO Advanced Research Workshop entitled "Recycling, Remediation, and Restoration Strategies for Contaminated Civilian and Military Sites in the Arctic Far North." The workshop was held in Kirkenes, Norway from 24 to 28 June, 1996. It was organized by the American Association for the Advancement of Science, Washington, DC, United States; the Norwegian Radiation Protection Authority, Oslo, Norway; and the Russian Research Center "Kurchatov Institute," Moscow, Russia.

The editor would like to thank Robert Dyer of the United States Environmental Protection Agency, Vitaly Lystsov of RRC "Kurchatov Institute," and Per Strand from the Norwegian Radiation Protection Authority for their hard work as workshop co-organizers. We also thank Dr. Luis Veiga da Cunha and the NATO Scientific Affairs Division for making this workshop possible, and the contributors to this volume for the time and effort they spent preparing the chapters.

Special gratitude is expressed to Sanoma Lee Kellogg of AAAS for formatting the book and for her part in editing the book for style and grammar.

Finally, we wish to thank all of the workshop participants for their contribution to the set of recommendations and comments included in the final chapter of this book.

I. INTRODUCTION

CHAPTER 1
CURRENT RADIOACTIVE CONTAMINATION ISSUES IN THE ARCTIC NORTH AND OPERATION AND INFRASTRUCTURE OF THE RUSSIAN NUCLEAR FLEET

VITALY N. LYSTSOV, NIKOLAI S. KHLOPKIN
RRC "Kurchatov Institute"
Moscow, Russia

1. Abstract

The experience gained through the operation of marine nuclear installations exceeds many thousands of reactor-years in terms of treating spent nuclear fuel, liquid radioactive wastes, and solid radioactive wastes. The environmental impacts of both normal operation and accidents are sufficiently well-known, and additional experience has been obtained in the course of operating and servicing the nuclear fleet. The source term for the environmental radioactive pollution of the Russian naval and merchant nuclear fleets needs to be assessed, as does their maintenance infrastructure. Different scenarios leading to the release of radionuclides into the environment should also be considered, after which it will be possible to define resulting effects on human health and the environment. The findings of these assessments can form the basis for prioritizing projects to remediate nuclear fleet contamination and improve the nuclear fleet maintenance infrastructure.

2. Introduction

The wide usage of marine nuclear installations in Russia began at the end of the 1950s and required the corresponding assessment of their impact on ship personnel and on populations living in the vicinity of naval and merchant bases, as well as on the environment as a whole.

The infrastructure for servicing the nuclear fleet was developed in parallel with fleet construction. It included shore technical bases (STBs) and floating technical bases (FTBs). Facilities for treatment and storage of liquid radioactive waste (LRW) and solid radioactive waste (SRW) and for temporary storage of spent nuclear fuel (SNF) were constructed at these bases and yards. However, serious problems of infrastructure functioning have accumulated, especially in the last ten years.

3

E.J. Kirk (ed.), Decommissioned Submarines in the Russian Northwest, 3–12.
© 1997 *Kluwer Academic Publishers.*

The federal goal-oriented program of the Russian Federation [1] on the treatment of radioactive waste states that about 33,000 cubic meters of LRW and SRW with a total activity of about 0.7 PBq (20 kCi) have accumulated so far at STBs and FTBs. The capacities of storage tanks and storage sites are practically exhausted, and the technical condition of these sites, which were built 20 to 30 years ago, remains unsafe. The greatest accumulated activity is connected with SNF storage sites. The largest portion of SNF is stored at the Andreev Bay STB in Zapadnaya Litsa fjord. A much smaller portion is stored at the STB in Gremikha, and the rest is kept on board FTBs [3]. More than 20,000 SNF assemblies were initially stored at Andreev Bay using wet storage technology, but after 1982, the storage facility's steel tanks became unstable and radioactive water began to leak into the environment. Authorities decided to unload the storage facility, and SNF assemblies were moved into a new, specially-constructed dry storage facility. The new SNF storage site at Andreev Bay was constructed only for temporary storage, and its prolonged use creates a situation of high potential risk. Potential risks also exist for some FTBs where SNF is kept. The most famous of these cases is the FTB *Lepse*.

In recent times, a new element of the nuclear fleet infrastructure—decommissioning—has emerged and is expanding very rapidly. The following objects are to be decommissioned:

- nuclear submarines;
- the nuclear icebreaker *Lenin;* and
- infrastructure vessels, such as the FTB *Lepse*.

More than 130 nuclear submarines are laid up [3] and need to be decommissioned. About two thirds of these submarines are stored afloat together with nuclear fuel that has not been unloaded, which strongly increases the risk of accidents. The fulfillment of the large-scale decommissioning program will lead to an increased accumulation of radioactive wastes and will seriously aggravate the unsolved problem of radioactive waste treatment. In any case, it is reasonable to expect that there will be additional environmental impacts from radioactive waste.

3. Radioactive Waste Production at Marine Nuclear Installations

There are two possible modes of radionuclide release to the environment from all of these objects:

- routine, controlled release; and
- accidental, uncontrolled release.

First, routine releases from typical marine nuclear installations will be considered. The latest installation, KLT-40, (used at icebreakers *Taymyr* and *Vaygach* and the cargo ship *Sevmorput*) will serve as an example. This installation has a nominal thermal capacity of 130 to 170 megawatts and is able to work without fuel reloading for three to four years [7]. As a result of the installation's operation, gaseous radioactive waste (GRW), LRW, SRW, and SNF are produced. GRWs are formed due to atmospheric argon activation and other noble gases released from the fuel. Recently, Kolomiets et al. [2] performed a comprehensive analysis of radioactive waste production at KLT-40. Annual production of radioactive noble gases by this installation varies between 0.04 and 0.4 TBq (one and ten Ci). SNF storage also produces ^{85}Kr in small quantities (not more than 0.4 TBq [ten Ci] for the entire storage period of the complete reactor core). Thus, the GRW problem for marine nuclear installations is insignificant. The same is true for ventilation of nuclear submarine reactor compartments in the normal mode of operation.

The greatest annual quantity of LRW (90 cubic meters) is formed during the year of fuel reloading. In other years, the figure may drop to ten cubic meters, but on average, it amounts to 30 cubic meters, with a total activity of about six GBq (0.15 Ci). The radionuclide composition of LRW is variable, with the most significant radionuclides being the fission products ^{137}Cs and ^{90}Sr and the activation products ^{60}Co and ^{54}Mn.

Most of the SRW (more than 70 percent by volume) belongs to the category of low-level waste. The high-level waste category makes up only five percent by volume, but 70 percent by activity level. The KLT-40 installation produces four cubic meters of solid radioactive waste with an activity of about 0.15 TBq (four Ci) each year. In addition, it produces 0.8 cubic meters of ion exchange resins with an activity of 7.5 to 180 GBq (0.2 to five Ci). In total, one year of marine nuclear installation operation produces about 35 cubic meters of radioactive waste by volume (mainly liquid) and 0.2 to 0.4 TBq (five to ten Ci) by activity (mainly in solid form).

4. Treatment of Radioactive Wastes

Most of the radioactive waste produced before 1992 was dumped at sea. The last discharge of LRW in the Far East was made by the Russian Navy in 1993. The sites, quantities, and activities of dumped wastes are thoroughly described in [5] and [6]. Liquid radioactive wastes are now treated only at the STB "Atomflot" in Murmansk. The pilot plant, which has been in existence there since 1986, is able to process only about 1000 cubic meters of LRW per year. That is enough for the needs of the merchant nuclear fleet, but quite insufficient for naval and decommissioning requirements. The capacity of the Murmansk facility should be increased by one order of magnitude, and a completely new facility for LRW treatment should be built in the Far East. International projects are now underway for both sites. Fulfillment of these projects will help to diminish the risk of accidental LRW leakages from storage tanks situated at STBs and FTBs. From a political standpoint, it is important that Russia will be able to join the ban on LRW discharges at sea accepted by members of the London Convention in 1972 once these treatment facilities are completed.

The existing LRW treatment facility in Murmansk operates through the following steps:

- electrochemical coagulation;
- mechanical filtration;
- ion exchange with the help of non-organic selective sorbents; and
- two-stage dialysis.

The purified water is released directly into Kola fjord after it is examined for radionuclide concentrations. Final concentrations must be lower than the Russian standards for drinking water, and actual levels of water contamination are significantly lower.

5. Radioactive Wastes from Decommissioning

Additional radioactive waste input into the environment is expected due to large-scale decommissioning processes. The main problem is the decommisioning of nuclear submarines, which begins with SNF removal from the ship reactor. As in the routine operation of nuclear fuel reloading, some gaseous-aerosol releases will take place and some LRW will be produced during this process. Additional LRW will appear after the drainage of primary and secondary circuits and decontamination. LRW output for one nuclear submarine is between ten and 100 cubic meters. Subsequent dismantling of the equipment produces up to ten cubic meters of SRW. The greatest problem is determining what to do with the reactor vessel and the reactor compartment. Currently, the most plausible solution seems to be to cut the reactor compartment with all of the equipment inside from the nuclear submarine hull, then seal it and store it somewhere for at least 30 years. Possible temporary storage sites include empty mining shafts, trenches in the permafrost at the archipelago Novaya Zemlya, and underwater grounds in shallow marine bays.

Which significant radionuclides could be found in such reactor compartment packages? In a reactor of the KLT-40 type that had functioned for 15 years, the following radionuclides would be present two years after reactor shutdown: ^{55}Fe, ^{60}Co, and ^{63}Ni with corresponding activities of about 1.8, 1.8, and 0.4 PBq (50, 50 and ten kCi). Ninety percent of these activities would accumulate in radiation screens and about ten percent in the material of the reactor vessel itself. Deposits on the surfaces of the primary circuit would add some fission products (^{90}Sr, ^{137}Cs) at a level of 0.4 TBq (ten Ci). After 30 years of reactor compartment storage, only about 0.6 PBq (15 kCi) of ^{63}Ni and 0.2 PBq (five kCi) of ^{60}Co would remain. These figures can be used as initial conditions in assessing the environmental impact of reactor compartment packages for different scenarios of their storage and dismantling.

The greatest potential for radioactive release into the environment is connected with operations involving SNF. Storage and transportation could be prone to accidents. SNF from one active zone of a KLT-40 reactor after three years of storage still has an activity of about 20 PBq (500 kCi), most of which is contributed by nearly equal portions of ^{90}Sr and ^{137}Cs. Spent nuclear fuel from the merchant nuclear fleet is stored mostly at the FTBs *Imandra, Lotta,* and *Lepse* in the Murmansk harbor. The *Lepse* contains some damaged SNF and is in rather poor technical condition. As was mentioned previously, SNF from naval operations is kept mostly at Andreev Bay, but also at Gremikha (15 percent) and FTBs (five percent) [3]. Planned transportation of SNF to the Ural Mountains is prevented mainly by economic difficulties. As a result, storage facilities are filled beyond current capacities, which increases the risk of accidents.

Information about the dumping of seven reactors with SNF and ten defuelled reactors into the Kara Sea during the period from 1965 to 1988 attracted significant international attention. Current verification of the radionuclide inventory in these reactors [4] gives a total activity for 1994 of 4.5 PBq (122 kCi), including 0.09 PBq (2.7 kCi) of actinides. The most urgent question now is how long protective barriers (including steel, bitumen, and furfural) will last in a marine environment and how quickly the leaching of radionuclides will proceed. Reactors without fuel could model reactor compartments from decommissioned submarines in the case of their disposal at underwater sites.

TABLE 1. Assessment of the Radioecological Significance of Various Problems of Operation of the Russian Nuclear Fleet

Type of Problem	Potential Radionuclide Release Level	Probability of Harmful Impact	Urgency of Remedial Measures	Expense Level	Priority Level
Unloaded SNF in floating nuclear submarines	very high	high	very high	very high	very high
Storage of SNF at naval facilities	very high	medium	high	high	high
Storage of SNF at the FTB *Lepse*	high	high	high	high	high
SRW storage at open temporary sites	medium	low	medium	medium	medium
LRW storage in tanks and insufficient treatment facilities	low	low	medium	low	medium
Reactor Compartment storage	medium	low	low	medium	low
Dumped radioactive waste and reactors in the Kara Sea	medium	low	low	high	low

This short review of radioecological problems connected with the operation of the Russian nuclear fleet is summarized in Table 1, where a priority choice for different problems is estimated. Here, the problems are categorized according to:

1. level of potential radionuclide release;
2. probability of harmful impact on human health and the environment;
3. urgency of remedial measures;
4. level of expenses in case such remedial measures are fulfilled; and
5. grade of priority choice (formulated by combining the previous four features).

Specific criteria need to be defined before a more precise assessment of the environmental impact of radioactive pollution can be determined.

The highest level of priority is given to the problem of unloaded SNF in nuclear submarines in floating storage. About two thirds of the nuclear submarines waiting to be decommissioned are moored with SNF on board. The total activity of this SNF is higher than the total activity of SNF at all storage facilities, both on shore and afloat. This is due both to the high quantity of SNF on board the submarines and to the shorter amount of time that has elapsed since reactor shutdown for fuel in submarines.

6. Safety Measures in the Decommissioning Process

The reactors of laid-up nuclear submarines have been safely shut down by reactivity control organs, and the following special measures have been taken to prevent the extraction of control rods from the active zone:

*   The electricity supply has been cut off (including excision of part of the feeding cable).

*   The control rod drives have been immobilized by welding.

*   Special rests have been attached to the shim rods to fix them to the floor.

Thus, the problems of nuclear safety can be considered to be solved for regular situations. However, one should also consider accidental situations such as ship collisions, falling aircraft, fires, explosions, and other potential dangers. Even if the probability of such events is low, the possibility of a criticality accident resulting in the case that one does occur must be planned for. The fact that the active zone of a reactor with no shim rods still has a significant excess reactivity even if the energy margins have been exhausted must be accounted for. Excess fuel reactivity may reach a level of five to ten beta, where beta is an effective share of delayed neutrons (3.5 to five percent). Thus, final safety of nuclear submarine reactors for all conditions can be provided only after SNF removal.

Once the SNF removal process is complete, the solution of radiation safety problems will also be simplified since more than 85 percent of the total activity is removed from the reactor of a nuclear submarine during that process. Any accident involving only the remaining activity will inevitably be much less significant.

The probability of a serious accident involving a nuclear submarine with fuel on board is proportional to the number of submarines in storage still containing fuel. As this number is presently rather large, the highest priority should be given to the program for unloading SNF from floating nuclear submarines.

7. Storage of Spent Nuclear Fuel

The second-highest priority should go to to the expansion of naval storage facilties, especially ashore. Existing storage facilities are overfilled and cannot accommodate additional SNF from submarines. The stipulated transportation of SNF to the reprocessing plant at the Mayak Association in the Ural Mountains has been proven inadequate. The number of special rail cars and TK-18 transport containers which correspond to the safety requirements of the International Atomic Energy Agency is not sufficient. In recent years, transport operations involving SNF have become prohibitively expensive. In addition, the volume of the storage facility at Mayak is insufficient. Lack of financing prevents prompt solutions to all these problems.

The most plausible solution for the nearest future is the construction of a new dry storage facility. Most of the SNF currently needing to be dealt with has been stored long enough that it has a low level of energy and radiation production. Consequently, cooling water is no longer a necessity and natural air circulation is sufficient to keep fuel cladding at the required temperature.

In comparison with wet storage, dry storage of SNF has significant advantages. In wet storage, the transfer of radionuclides into cooling water occurs due to fuel corrosion. Leakage of this water may lead to environmetal contamination, as was the case with the storage facility at Andreev Bay. The presence of a moderator (i.e. water) in a storage facility increases the breeding ratio, which necessitates additional precautions. Wet storage is more expensive than dry storage, and it cannot be considered a desirable methodology for the construction of future facilities. The relatively lower cost of constructing dry storage facilities makes the program of expansion of naval storage facilities to accomodate SNF from submarines realistic, and this program is, accordingly, assigned second-highest priority in Table 1. At a new dry storage facility, the problem of corrosion will be absent, degradation of fuel and equipment will be minimal, and operational and control costs will be low.

By applying new technological decisions, design of the dry storage facility can be significantly simplified. The fuel can be stored in the same containers used for transportation. Proposals for the construction of suitable containers have been put forth by several United States firms (e.g. INEL and NAC) and by the Russian design bureau KBSM, a St. Petersburg-based firm that developed a new design for a reinforced concrete container. Using such containers will make it possible to develop new schemes for SNF transport and storage. It will enable spent nuclear fuel from floating submarines to be unloaded directly into transport containers, which can then be kept at storage facility grounds for a sufficiently long time without reloading. Later, the SNF in these containers can be sent to a reprocessing plant or to the site of final disposal.

Leaving SNF in storage on board the submarine for a long period after reactor shutdown results in a lower level of residual energy production in the SNF, which makes it possible to unload SNF from the submarine directly at the pier, thereby eliminating the necessity of intermediate storage at FTBs. The safety of this operation can be increased by draining the reactor (removing the moderator), which simplifies safety provisions and reduces the possibility of an uncontrolled chain reaction occurring as the result of equipment failures or human error.

At present, there is a shortage of reloading equipment. However, new equipment can be made in much lighter and simplified versions since the SNF in the submarines has lower levels of energy and radiation production due to the amount of time that has passed while the vessels have been awaiting decommissioning.

8. The *Lepse*

New means and equipment will be required to extract damaged and normal SNF from the FTB *Lepse*. Despite its lower potential for radionuclide release, this program received high priority level in Table 1 because the *Lepse* is situated in the harbor of densely-populated Murmansk, which increases the probability of harmful impact and, thus, the urgency of remedial measures.

9. Waste Treatment

The problems of solid and liquid radioactive waste treatment were assigned medium priority level in Table 1 mostly due to the lower total activities involved. However, treatment facilities for LRW and SRW are usually close to living areas, so the solution of these problems has high social priority. In order to solve these problems, local interim storage sites need to be defined.

10. Reactor Compartments

The storage of reactor compartments obtained in the nuclear submarine decommissioning process was assigned low priority in Table 1. This is because the process of cutting out the reactor compartment has only recently begun and good isolation of radionuclides inside reactor compartments strongly reduces the probability of harmful impacts on human health and the environment.

11. Kara Sea

The problem of dealing with dumped radioactive waste and reactors in the Kara Sea was also assigned low priority since recent international studies indicate that the possible environmental effects from these objects are rather limited. At the same time, it seems that any attempt at remedial measures for these objects will be rather costly.

12. Conclusion

The authors of this paper maintain that the problems connected with the unloading of fuel from submarines and the storage of SNF should have the highest priority in allocation of both effort and finance.

References

1. Federal Goal-oriented Program of the Russian Federation on the Treatment, Utilization, and Disposal of Radioactive Waste and Spent Nuclear Materials for the years 1996 to 2005 (11 November 1995) *Rossiyskaya Gazeta*.

2. Kolomiets, B.I., Samorodov, A.F., and Filippov, M.P. (1993) Radioactive Wastes of Nuclear Icebreakers, in *Fourth Annual Science and Technical Conference of theNuclear Society of the USSR*, Nuclear Society of Russia, N. Novgorod, 857-859 (in Russian).

3. Petrov, O. (1995) Radioactive Waste and Spent Nuclear Fuel in the Navy of Russia, in P. Strand, Ed., *Environmental Radioactivity in the Arctic: Proceedings of the Second International Conference, Oslo, Norway 21-25 August 1995*, Osteras, 404-406.

4. Sivintsev, Yu. and Kiknadze, O. (1995) Radioecological Danger of Long-lived Radionuclides in Nuclear Reactors Dumped in the Arctic, in P. Strand, Ed., *Environmental Radioactivity in the Arctic: Proceedings of the Second International Conference, Oslo, Norway 21-25 August 1995*, Osteras, 152-155.

5. Sivintsev, Yu. (1994) *The Study of Nuclide Composition and Fuel Characteristics in Dumped Submarine Reactors and the Atomic Icebreaker* Lenin, Part I, Atomic Icebreaker, IAEA-IASAP-1.

6. Sivintsev, Yu. (1994) *The Study of Nuclide Composition and Fuel Characteristics in Dumped Submarine Reactors and the Atomic Icebreaker* Lenin, Part II, Atomic Icebreaker, IAEA-IASAP-1.

7. Vasyukov, V.I., Starodubtsev, V.G., Petrov, T.T., Melnikov, E.M., Polunichev, V.I., Mitenkov, F.M., Samoilov, O.B., and Panov, Yu.K. (1990) Experience of Development and Operation of NSS for Civilian Vessels, in B.K. Yasnevsky (ed.), *Atomic Energy at Sea: Proceedings of an International Scientific Seminar, Murmansk, Russia, 24-28 September 1990*, Murmansk, 47-52.

CHAPTER 2
WORLDWIDE DECOMMISSIONING OF NUCLEAR SUBMARINES

Plans and Problems

POVL L. ØLGAARD
Reactor Safety Program, Risø National Laboratory
Department of Physics, Technical University of Denmark
Lyngby, Denmark

1. Abstract

The paper begins with a review of the magnitude of the task of nuclear submarine decommissioning. Next, the various stages of nuclear submarine decommissioning are discussed together with a review of the procedures selected by Russia, the United States, the United Kingdom, and France. The nuclear risks connected to decommissioning are also considered, followed by a comparison of the three options for final disposal of the major radioactive components of the reactor compartment—sea disposal, shallow land burial, and deep land burial. Finally, the paper addresses the special problem of nuclear submarines with damaged cores.

2. Introduction

Since the 1950s, a number of countries—the United States, the Soviet Union/Russia, the United Kingdom, France, and China—have built nuclear propelled vessels, primarily submarines. The concentration has been on submarines because it was the introduction of the nuclear reactor as a propulsion source that permitted the realization of the genuine submarine, i.e. a vessel that can sail fully submerged at full speed for practically unlimited periods of time.

Another important sea application of nuclear power is the use of reactors for propulsion systems in Arctic icebreakers. Because of its long Arctic coastline and the need for transport to and from coastal towns, Russia has built a number of nuclear icebreakers which can sail for years with no need for bunkering.

A limited number of other nuclear vessels have been built. The United States has built nuclear aircraft carriers and cruisers, Russia has built nuclear cruisers, and France is building a nuclear aircraft carrier. In addition, the United States has built a combined

E.J. Kirk (ed.), Decommissioned Submarines in the Russian Northwest, 13–23.
© *1997 Kluwer Academic Publishers.*

cargo and passenger ship, Germany an ore carrier, and Japan an oceanographic research vessel. These three civilian ships were each taken out of service after a few years of operation.

Nuclear vessels, like ordinary vessels, do not last forever. Usually they have a lifetime of about 25 years. Since the construction of nuclear submarines started in the mid-1950s, a number of submarines have already been taken out of service and are in the process of being destroyed. Since the submarine dominates the field of nuclear vessels, the following considerations on decommissioning will be limited to submarines, but the approaches discussed apply to other nuclear vessels, as well.

3. The Magnitude of the Problem

The magnitude of the problem is illustrated in Table 1, where the total number of nuclear ships built are listed together with the total number of nuclear submarines built, decommissioned, and in operation.

TABLE 1. Nuclear-Propelled Vessels and Submarines 1995

	Vessels built	Submarines built	Submarines decommissioned	Submarines operational
USA	199	182	82	100
Russia	257	244	124	120
UK	25	25	9	16
France	12	12	1	11
China	6	6	0	6
Japan	1	0	0	0
Germany	1	0	0	0
Total	501	469	216	253

Russia has built more nuclear vessels than all other countries combined, and the nuclear submarine is the dominating nuclear vessel type. Almost half of the nuclear vessels built have, by now, reached the decommissioning stage.

The figures given in Table 1 may not be completely correct. However, when figures from different sources are compared, the disagreement is usually quite small.

Since most of the Russian nuclear submarines have two reactors while those of the other countries in almost all cases have one, the total number of reactors involved in nuclear submarines is about 700. This figure is sometimes compared to the number of nuclear power reactors in the world. At the end of 1994, 432 power reactors were in operation and about 70 had been shut down. However, such a comparison is misleading since the thermal power level of modern nuclear power reactors is usually around 3000 megawatts (MW) while that of submarine reactors is usually around 100 MW. Thus, the amount of radioactivity contained in a naval reactor is usually a factor of approximately 30 lower than that in a modern power reactor, and the potential release of radioactivity is much smaller for a submarine reactor.

With a few exceptions, which will be disregarded here, submarine reactors are all so-called pressurized water reactors which use enriched uranium (20 to 90 percent ^{235}U) as fuel.

4. Stages in Nuclear Submarine Decommissioning

The nuclear submarine decommissioning process may be divided into inactivation and scrapping.

The first step in inactivation is to remove all weapons from the ship. This usually takes place at the home base.

After the submarine has moved to the defuelling facility, the reactor is shut down for the last time and allowed to cool down. During the cooling period, expendable and sensitive materials and the main batteries are removed, liquid and gas circuits and tanks are emptied and dried out, and the electrical systems are de-energized.

Next, the submarine hull above the reactor is removed and the reactor is defuelled, i.e. the spent fuel is moved in shielding containers from the reactor to an intermediate storage facility, and the reactor circuits are emptied and dried out. Initially, the spent fuel is stored in a pool (wet storage), but after a number of years it may be transferred to dry storage facilities where the reduced heat production is removed by the natural circulation of air. Ultimately, the fuel may be sent to a chemical reprocessing facility, where the remaining uranium may be extracted for recycling, or the spent fuel may be considered high-level waste and disposed of in an underground depository.

Once the fuel has been removed, the radioactivity of the submarine is reduced by a factor of approximately 100. The remaining radioactivity is due to neutron activation of the reactor tank, its internals, the radiation shield around the reactor, and possibly also the steam generators and pumps. Corrosion products in the reactor circuit also play a role. All of these components are situated in the reactor compartment. In most cases, the next step is to cut the reactor compartment out of the submarine and seal the compartment at both ends.

The scrapping phase involves cutting up the remaining part of the submarine and re-cycling the materials to the extent possible. Nuclear submarines contain significant amounts of hazardous and toxic materials such as lead, PCBs, and asbestos, which must be handled properly when the submarine is cut up.

The procedure used in decommissioning nuclear submarines is, in general, quite similar for various countries. However, the final handling of the reactor compartment varies, and some countries have special problems. For these reasons, it is worthwhile to consider the national differences.

4.1. UNITED STATES

At present, all United States nuclear submarines are decommissioned at Puget Sound Naval Shipyard near Seattle, Washington. The defuelling is performed in a dry-dock. Formerly, the spent fuel was sent to a reprocessing plant, but now it is considered high-level waste and will be disposed of in a geological repository if and when such a facility becomes available.

Next, the reactor compartment is cut out of the submarine and sealed at both ends. It is placed on a barge and transported along the Washington coast down to the Colombia River and up the river to the Hanford Reservation. Here the reactor compartments are placed in a trench which will ultimately be filled in. The reactor compartment typically has a diameter and length of about ten meters, a volume of about 1000 m^3, and a mass of about 1000 tons.

The Hanford Reservation is situated in a semi-arid region with little precipitation. It is the site of the Hanford production reactors, which are now closed. The site has been used for shallow land burial of other radioactive wastes for a number of years.

The United States has the capacity to dispose of about five nuclear submarines per year. The cost of inactivating and scrapping a nuclear submarine is US$35 to 40 million.

4.2. RUSSIA

Russia, too, has started the decommissioning and disposal of nuclear submarines. However, in some respects the procedures used differ from those used in the United States. Defuelling is often performed at sea using floating cranes, and the spent fuel may be stored in special storage ships for extended periods of time. The reactor compartment may be cut out and the remaining parts of the submarine scrapped at various shipyards, e.g. the Zvyozdochka plant in Severodvinsk, the Nerpa plant near Murmansk, and the Zvezda plant in the Far East.

The economic difficulties of Russia have strongly affected the inactivation and scrapping of decommissioned Russian submarines.

One problem is that the interim storage facilities for spent fuel of both the Northern and the Pacific fleets are full and, at the same time, there have not been sufficient funds to maintain the facilities, whether land or ship storage. Formerly, the spent fuel was sent to the Mayak reprocessing plant after an interim storage period at the naval bases. Today, it seems that the Russian Navy does not have the funds to pay for the reprocessing and that the Russian nuclear safety authorities will not approve the existing transport containers for long-distance transport. A consequence is that many of the decommissioned nuclear submarines have not been defuelled, but rest moored at naval bases. An estimate of the state of the Northern Fleet submarines is given in Table 2.

TABLE 2. Status of the Nuclear Submarines of the Northern Fleet, 1995

Defuelled submarines	18
Decommissioned submarines with fuel	52
Decommissioned submarines	70
Operational submarines	81
Total number of submarines	151

Again, these figures may not be quite up-to-date, but the error is believed to be small. The 18 defuelled submarines are in different stages of inactivation or scrapping. It is estimated that the reactor compartment of about ten of the defuelled submarines has been cut out.

Russia does not have a repository for the reactor compartments similar to the United States repository at the Hanford Reservation. Earlier, the Soviet Union disposed of reactor compartments by sinking them at sea. The Northern Fleet used sites close to Novaya Zemlya. However, in 1976 the Soviet Union became party to the London Convention, which prohibits sea disposal of high-level radioactive waste, so this option is no longer available. Therefore, the reactor compartments are left floating at various naval facilities until a decision on the future of the compartments has been made.

Initially, the cut-out reactor compartments were joined to pontoons to ensure the necessary buoyancy during floating storage. Later, so-called three-compartment units were preferred. This solution involves cutting out the reactor compartment together with the two adjoining compartments. The heavy components of the two adjoining compartments are removed and their ends are sealed. The adjoining compartments then ensure the necessary buoyancy of the unit during floating storage.

In Russia, various possibilities for the future handling of the reactor compartment are being considered. Many organizations involved seem to prefer an extended storage period (70 to 100 years) for the compartment and subsequent reuse of most of the materials. Whether such reuse will be economical is questionable. The extended storage could take place in underground caves at the Kola peninsula, in permafrost areas at Novaya Zemlya, in trenches on shore, at open sites, and in coastal water areas. No decision has yet been made on which extended storage method should be used.

4.3. THE UNITED KINGDOM AND FRANCE

As shown in Table 1, France and the United Kingdom have a much smaller nuclear fleet, and they do not have access to burial sites similar to the Hanford Reservation. However, both countries have established organizations—ANDRA in France and Nirex in the United Kingdom—which will be responsible for disposing of the radioactive waste, civilian or military, which is produced in these countries. Both organizations are preparing a deep burial repository.

The shafts down to such deep repositories have limited dimensions for economic reasons. Nirex will only allow packages with dimensions of 1.7 meters by 1.7 meters by 1.2 meters. Thus, the reactor compartments will have to be cut into smaller pieces before disposal. For this reason, the United Kingdom and France store the inactivated nuclear submarines for a period of 15 to 20 years or more before dismantling the reactor compartments. This permits the reduction of the radioactivity of the reactor component by a factor of ten or more, which also reduces the radiation dose to the personnel that will cut and package the radioactive parts. In France, the reactor compartment is cut out of the nuclear submarines and stored under dry conditions at the Direction des Constructions Navales (DCN) in Cherbourg. In the United Kingdom, the inactivated submarines are left floating at a naval base under constant supervision.

5. Radiological Risks Connected to Nuclear Submarine Decommissioning

It has sometimes been claimed that accidents with nuclear submarines, including decommissioned submarines which have not been defuelled, could have consequences similar to those of the Chernobyl accident. This is not the case for two reasons. First, as mentioned above, the power level of the submarine reactors is about 30 times smaller than that of the Chernobyl reactor, and the content of radioactive materials is also around 30 times lower. Therefore, the potential release of radioactivity is much smaller. Second, almost all nuclear submarines are provided with pressurized water reactors, which have significantly better safety properties. This does not mean that accidents with decommissioned submarines may not occur, but if they occur, the consequences will be much more localized.

A number of different types of nuclear accidents exist. They are discussed briefly below.

5.1. REACTIVITY ACCIDENTS

Reactivity accidents involve a runaway chain reaction in a reactor. This type of accident has occurred in connection with refueling nuclear submarines, e.g. the Chazhma Bay accident near Vladivostok in 1985. No report is available of reactivity accidents in decommissioned reactors, but they can not be excluded. They may occur in connection with defuelling. If, for example, the control rods have not been detached from the reactor vessel lid before the lid is lifted, the reactor will go supercritical. The energy produced in the power excursion will probably be less than in the Chazhma accident since the excess reactivity is likely to be lower in a reactor just before defuelling than in a reactor with fresh fuel. But the content of radioactivity in the burned-out core is much higher. Therefore, the effect of such an accident is likely to be more localized than in the case of the Chazhma accident, but the local contamination will probably be significantly higher.

Needless to say, reactivity accidents may only occur when there is still fuel in the reactor. Also, if the defuelling is performed in accordance with regulations, no accident will occur.

Reactivity accidents might also occur in interim fuel storage facilities if the stored fuel elements are moved closer together to increase the storage capacity of the facility. However, this type of accident is unlikely since only "safe geometries" are authorized for such facilities, and because fuel elements are usually moved one by one so that the maximum excess reactivity, should it materialize, will be quite small.

5.2. LOSS-OF-COOLANT ACCIDENTS

Loss-of-coolant accidents (LOCAs) have occurred in operational submarine reactors, e.g. the Echo II accident near Bear Island in 1989, but no information is available on such accidents in decommissioned submarines. LOCAs occur if the coolant water drains out of the reactor due to a leakage in the reactor circuit and the fuel elements are no longer cooled. Thus, their temperature increases, and the fuel may melt and release radioactivity. LOCAs occur primarily in nuclear submarines during operation, but they may also occur in decommissioned submarines that have not been defuelled. The reason is that even after shutdown, the reactor fuel will produce heat due to radioactive decay of the fission products, and this heat has to be removed. However, after a shutdown period of two to three years, the decay heat will no longer be strong enough to damage the fuel elements, even if all water is drained out of the reactor.

LOCAs might also occur in pool storage facilities for spent fuel if water is drained out of the pool and the cooling time of the fuel has been sufficiently short. However, drainage of such a pool should be much slower and the possibility of supplying replacement water should be much better than in the case of a nuclear submarine.

5.3. LOSS OF WATER SHIELDING

The water in a storage pool for spent fuel has two purposes: it acts as a coolant for the fuel to remove the decay heat, and it acts as a radiation shield for the gamma radiation emitted from the fuel elements. If the water is drained out of the pool, not only the cooling, but also the shielding, is lost. The removal of the shield would result in a strong radiation field close to the pool, but the effect would be quite local. Further, it should be fairly easy to replace any leaking water by filling extra water into the pool. According to Western sources, there has been at least one case when significant amounts of water leaked out of a spent fuel storage pool of the Northern Fleet, but not to the extent that the water shield disappeared.

5.4. LEAKS OF RADIOACTIVITY

A number of possibilities exist for leakages of radioactive materials from decommissioned submarines.

As long as a nuclear submarine has not been defuelled, the reactor coolant—which is slightly radioactive—may leak out of the reactor circuit and possibly out of the submarine. However, the subsequent contamination should be quite limited since the reactor water is presumably cleaned continuously.

Once the submarine has been defuelled and the reactor circuit has been dried out, all radioactive materials present in the submarine are solid. Even if the submarine or the reactor compartment sank and water got into the compartment, the release of radioactivity to the environment would be quite slow since corrosion of steel is slow. In addition, it should not be too difficult to salvage the submarine or compartment rapidly after the sinking.

Spent fuel storage facilities of the pool type may give rise to water leakage as discussed in section 5.3. The consequences of such a leak are not too serious provided the water purity is kept high by continuously cleaning the pool water. Western sources indicate that the purity of the water in the spent fuel storage pools of the Northern Fleet is not always up to Western standards.

Nuclear submarines with damaged cores represent a special risk for contamination. They will be discussed in more detail in section 7.

The main risks of nuclear accidents are connected to the reactor fuel. Once the fuel has been sent away, the risk of radiological accidents is greatly reduced.

6. Comparison of Reactor Compartment Disposal Methods

As mentioned in section 4.1, 4.2 and 4.3, various methods have been used, are used, or will be used for reactor compartment disposal.

6.1. SEA DISPOSAL

In 1959, the United States Navy disposed of the first reactor vessel (without fuel) of the NSS Seawolf submarine, dumping it in the Atlantic Ocean. Since then, the United States has not used sea disposal. The Soviet Union used sea disposal of damaged reactor compartments from 1965 to 1988. The dumping took place as mentioned above in the sea near Novaya Zemlya. Four reactor compartments with three damaged cores, one complete submarine with two reactors containing fuel, and three reactor tanks (one with fuel) were dumped.

Sea disposal has a number of advantages. It is cheap and results in low radiation doses to the personnel involved. Further experience has shown that the release of radioactivity to the environment has, so far, been very low. However, the main and prohibiting disadvantage of sea disposal is that it has been banned by the London Convention.

6.2. SHALLOW LAND BURIAL

The United States uses shallow burial for the reactor compartments at the Hanford Reservation, a semi-arid region. Even though Russia has not yet decided on the disposal method, it seems that Russia will also use some form of shallow land burial (cf. 4.2).

This method has the advantages that the radiation dose to the personnel involved is low and the release of radioactivity to the environment is slow and can be monitored. However, it is more expensive than sea disposal, and the disposal site requires long-term surveillance.

6.3. DEEP LAND BURIAL

As mentioned in section 4.3, the United Kingdom and France plan to use deep land burial for their nuclear submarines. Since, in this case, it is necessary to cut the reactor compartment and its components into pieces, the compartments or the submarines are left for the radioactivity to decay for a number of years.

The method is very expensive, and it will unavoidably give a considerably higher radiation dose to the personnel involved, even if the scrapping of the reactor compartment takes place 20 years after decommissioning.

7. Nuclear Submarines with Damaged Cores

Russia has a special problem since five of its decommissioned submarines contain damaged fuel that can not be taken out of the reactor. Two of these submarines belong to the Northern Fleet and three to the Pacific Fleet. The two Northern Fleet submarines seem to be an Alfa and an Echo II submarine which have both suffered a loss-of-coolant accident. The reactor compartment of the Alfa submarine has been cut out of the submarine as a three-compartment unit. One of the three Pacific Fleet submarines is the Echo II submarine that suffered the reactivity accident at Chazhma; the other two have suffered LOCAs.

Further handling of these submarines, which are strongly contaminated with radioactive materials due to the accidents, will be very complicated and expensive if the fuel has to be taken out, and the personnel involved in such an operation will unavoidably receive significant radiation doses. Should one of the submarines sink during floating storage, significant contamination of the areas nearby could result. Earlier, the Soviet Navy used sea disposal in such cases, but this disposal method is prohibited by the London Convention.

Considering this problem, it might be reasonable to take the following factors into account:

- Russia has many environmental problems, the solution of which will demand huge resources.

- Russia has economic problems which are likely to last for years.

- The West will help Russia, but help only. Russia will have to pay most of the bill.

- Therefore, Russia will have to choose affordable solutions, not the ideal solutions.

- Experience with reactor compartments dumped near Novaya Zemlya has so far shown that the release of radioactivity is small and slow.

- The total radiation dose received by human beings is likely to be much larger in cases where the fuel has to be cut out of the reactor than in cases where the reactor compartment or the entire submarine with fuel is dumped at a proper place in the ocean.

This raises the question:

Considering all these factors, would it not be to the advantage of everybody to make an exception to the London Convention and allow Russia to use sea disposal for the five submarines with damaged fuel?

The further details of how and where the dumping could take place should, of course, be negotiated.

References

1. LeSage, L.G. and Sarkisov, A.A. (in press) *Nuclear Submarine Decommissioning and Related Problems*, Kluwer Academic Publishers, Dordrecht.

2. Ølgaard, P.L. (1993) Decommissioning of Naval Nuclear Ships, NT-6, Technical University of Denmark.

3. Ølgaard, P.L. (1994) Potential Risks of Nuclear Ships, NT-11, Technical University of Denmark.

4. Ølgaard, P.L. (1996) Potential Sources of Cross-Border Radioactive Pollution due to Defense-Related Activities, Risø-I-1020(EN), Risø National Laboratory.

5. Semenov, Yu.P. (1995) Program of Complete Disposal of Russian Nuclear-Powered Submarines Decommissioned from the Northern Fleet, Paper prepared to support implementation of agreement between Kværner Moss a.s. and S.P. Korolets Rocket and Space Corporation Energia.

6. Yablokov, A.V. et al. (1993) *Facts and Problems Related to Radioactive Waste Disposal in Seas Adjacent to the Territory of the Russian Federation*, Office of the President of the Russian Federation.

7. Defense Committee, House of Commons. (1989) *Decommissioning of Nuclear Submarines*, Her Majesty's Stationery Office.

8. The International Institute for Strategic Studies. (1993) The Military Balance 1993/94, 1994/95, 1995/96, *U.S. Naval Nuclear Powered Submarine Inactivation, Disposal and Recycling*, United States Department of the Navy.

II. SPENT NUCLEAR FUEL

CHAPTER 3
THE DEVELOPMENT OF A COMPREHENSIVE UNDERSTANDING OF THE HANDLING AND TRANSPORT OF SUBMARINES' SPENT FUEL OUT OF NORTHWESTERN RUSSIA

VALERIJ ABRAMUSHKIN
RKK Energia
Moscow, Russia
BJØRN BORGAAS, KÅRE RYGG JOHNSEN
Kværner Maritime a.s.
Lysaker, Norway

1. Introduction

The company Kværner Maritime and the Russian Rocket and Space Corporation Energia have initiated a joint effort aiming to establish a basis for decisions on the subject of how to best handle the radioactive threat from decommissioned nuclear submarines laid up in waters close to the Norwegian border.

The Norwegian motivation for involving itself heavily in these questions was presented in the *Stortingsmelding nr. 34 (1993-94)*, a communication from the Ministry of Foreign Affairs to the Norwegian parliament detailing the Norwegian concerns with the hazards from nuclear activities in the Northern region. In brief, the motivation can be summarized as follows:

- There is a small, but not insignificant, possibility that certain accidents will expose parts of the Norwegian population to higher than normal doses of ionizing radiation.

- The same accidents may result in radioactive contamination of the environment and subsequent contamination of the food chain.

- Furthermore, enhanced levels of radionuclide contamination of Arctic waters as a consequence of accidents or long-term leaks will have a detrimental effect upon Norwegian commercial fisheries and fish exports.

A group of senior officials was established as a follow-up to this communication. A representative from the Ministry of Foreign Affairs heads the group, and the Ministries of Environment, Defense, Health, Fisheries, and Finance are represented, as is

E.J. Kirk (ed.), Decommissioned Submarines in the Russian Northwest, 27–40.

the Radiation Protection Authority. All work performed by the joint Kværner Maritime/RRK Energia project has been done for, and reported to, this group.

The work began with high-ranking representatives of the Russian Federation Navy, the Ministry for the Defense Industries, and the Ministry for Atomic Energy cooperating in the development of an overall Russian description of the status and plans for future actions regarding decommissioned submarines. Other Russian institutions involved in the subject in various ways also participated. This description was submitted to the Norwegian group of senior officials.

Kværner/Energia then evaluated the description and presented recommendations of future actions, all aimed at solving the environmental problems safely and efficiently.

Based on Kværner/Energia recommendations, Norway and Russia wrote a protocol stating an intent to reach an Agreement on their implementation.

1.1. WHAT TO DO

Almost all radioactive materials in a nuclear-powered submarine are found in its nuclear fuel. The major concerns are, therefore, related to the handling and storage of spent nuclear fuel (SNF). The SNF can be retrieved from the reactors and transported out of the region as a separate operation, independent of the other disposal and scrapping work. The costs of this operation are only a small fraction of those of the overall disposal operation.

By decommissioned submarines, the Kværner/Energia project includes only submarines which have had their weapons systems removed. Hazards from nuclear weapons are not a part of the scenario under study.

The recommendations in the protocol are:

1. There are no adequate means of seaborne transportation for spent fuel along the coasts of the Kola peninsula. A special carrier should be built.

2. Railway transport capacity out of the region is too small to achieve the fuel handling rate set by the Russian side. More special railway wagons are needed.

3. The NPO Zvyozdochka yard in Severodvinsk should receive support to upgrade storage tanks for liquid radioactive waste (LRW).

4. A mobile plant for volume reduction of liquid waste should be purchased.

5. A plan for and efforts aimed at emptying the Andreev Bay SNF storage site should be initiated.

6. The possibilities for Norwegian participation in the building of a storage facility for SNF in Mayak should be investigated.

7. The possibilities for Norwegian participation in the building of a storage facility for solid radioactive waste in the region should be investigated.

2. Overall Situation

Responsibility for a decommissioned submarine lies with the Naval unit to which it belonged during its operational service. There is no centralized and dedicated organization bearing overall responsibility in these matters.

Today, a total of about 90 nuclear submarines has been taken out of service from the Northern Fleet. They are laid up in the Kola/Severodvinsk area. The nuclear fuel remains in about 70 of these submarines. There are plans to remove another 35 nuclear submarines from service by 2010.

An unknown number of reactors in these submarines contain damaged fuel. Estimates of the number vary from two percent to roughly ten percent of the reactors. Neither information about the nature of the damage, nor the consequences for later fuel-related operations, are available.

From the Russian side, four yards are dedicated to the scrapping work:

• NPO Zvyozdochka, Severodvinsk;
• SRS Nerpa, Snezhnogorsk;
• PO Zevmash, Severodvinsk; and
• Naval Shipyard No. 10, Polyarny.

Together, they have a stated total scrapping capacity of three to four submarines per year. This is about half of what will be needed to reach the goal of 125 submarines by 2010.

The spent nuclear fuel is planned to be recycled into the overall Russian nuclear fuel cycle, which means it will be transported to the Mayak plant in the southern Urals for reprocessing.

The reactors themselves represent the largest single source of other radioactive wastes. A transport system and a tunnel-based long-term storage system have been described by the Russian side.

Figure 1 shows the plans presented by the Russian side displaying all investments necessary for complete scrapping of the submarines.

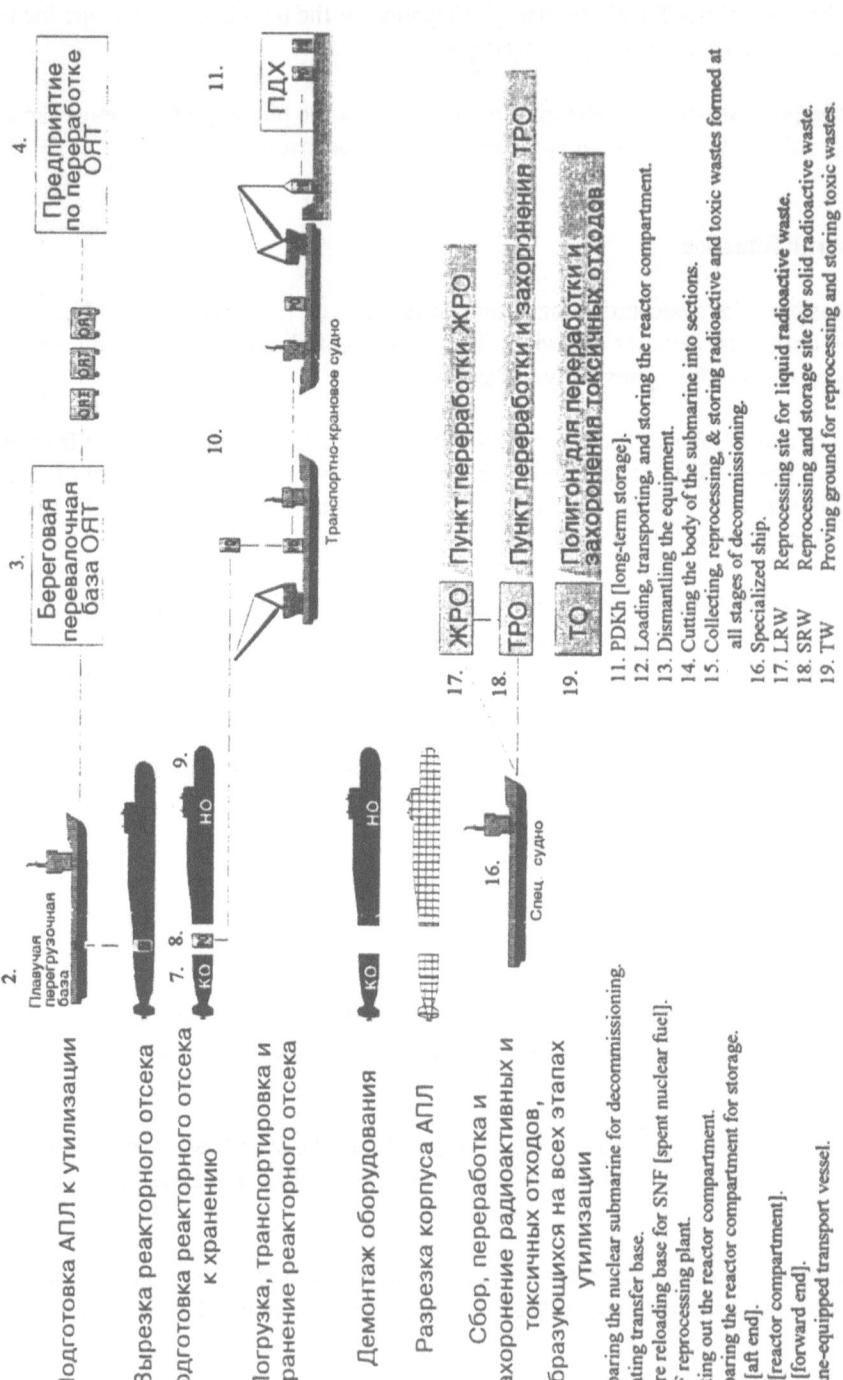

1. Подготовка АПЛ к утилизации
5. Вырезка реакторного отсека
6. Подготовка реакторного отсека к хранению
12. Погрузка, транспортировка и хранение реакторного отсека
13. Демонтаж оборудования
14. Разрезка корпуса АПЛ
15. Сбор, переработка и захоронение радиоактивных и токсичных отходов, образующихся на всех этапах утилизации

1. Preparing the nuclear submarine for decommissioning.
2. Floating transfer base.
3. Shore reloading base for SNF [spent nuclear fuel].
4. SNF reprocessing plant.
5. Cutting out the reactor compartment.
6. Preparing the reactor compartment for storage.
7. KO [aft end].
8. RO [reactor compartment].
9. NO [forward end].
10. Crane-equipped transport vessel.
11. PDKh [long-term storage].
12. Loading, transporting, and storing the reactor compartment.
13. Dismantling the equipment.
14. Cutting the body of the submarine into sections.
15. Collecting, reprocessing, & storing radioactive and toxic wastes formed at all stages of decommissioning.
16. Specialized ship.
17. LRW   Reprocessing site for liquid radioactive waste.
18. SRW   Reprocessing and storage site for solid radioactive waste.
19. TW   Proving ground for reprocessing and storing toxic wastes.

Figure 1. Diagram of the Decommissioning of a Nuclear Submarine

Considering the immediate Norwegian concerns and the limited funding available, it became an important part of the work of the Kværner/Energia group to understand the priorities. For that purpose, the group broke down the overall operation into 37 independent phases. For each operation, Kværner/Energia estimated resource and site requirements. It became clear that the SNF handling can be considered as independent of the rest of the disposal operation.

The group now believes that it has an essentially correct picture of how many submarines of which class are laid up where. It cannot, however, apply this information to any in-depth planning of the various operations, nor has this been its task. Much of the information about the submarines and their reactors is classified.

Figure 2 is based on data published in *Militbærbalansen*. The correlation of age to fraction in service was performed by Kværner/Energia. This figure shows that, with minor exceptions, submarines are taken out of service primarily as a result of their age. This analysis provides a general picture speaking against any generic failure modes which might have had an influence on the planning of the disposal operation.

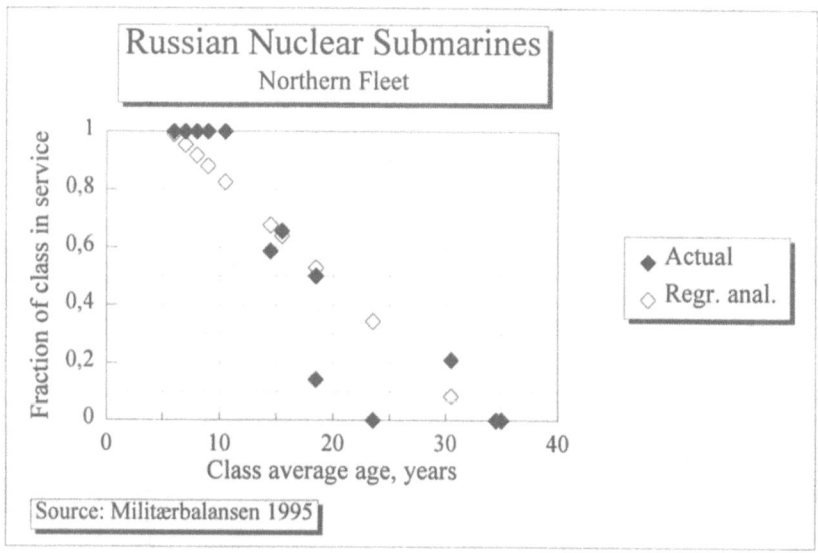

Actual = Actual data for each class of submarines
Regr. anal. = Linear regression analysis based on actual data for each class of submarines

Figure 2. Correlation of Submarine Age to the Fraction of Each Class Remaining in Service

Fuel retrievability is the single most important question to clarify in planning the disposal process. A reactor with fuel which cannot be retrieved represents a problem for which there are, at present, no solutions. The situation regarding fuel retrievability is somewhat unclear. This project has been informed about two submarines with unretrievable fuel. One is a Papa-class lying in Severodvinsk. The other is one of the

Charlie or Echo type submarines lying in the Ura bay[1]. However, other sources claim higher figures, up to eight.

As for other aspects relating to the condition of the submarines, there are no indications that they will cause problems for the disposal process, structurally or otherwise.

The disposal rate given by the Russian side, and reflected in the Figure 3, was used in the Kværner/Energia evaluation. In brief, decommissioning is foreseen to gradually decrease to a rate of three submarines per year for the years 2000 to 2010. Defuelling is anticipated to remain at around three to four vessels per year until 2000, when it will increase to twelve per year for a four-year period. In the last years of the planning period, the defuelling rate will again be about three vessels per year. The most significant trait of the plan is the making of three-compartment units as a means of intermediate storage for reactor sections. They increase in number up to 42 in 2000. After 2000, when completion of the reactor storage is anticipated, the reactor sections will be prepared directly for dry storage. In the peak years 2002 to 2005, roughly 18 reactor sections will be made, either from the three-compartment units or directly from intact submarines, and stored each year.

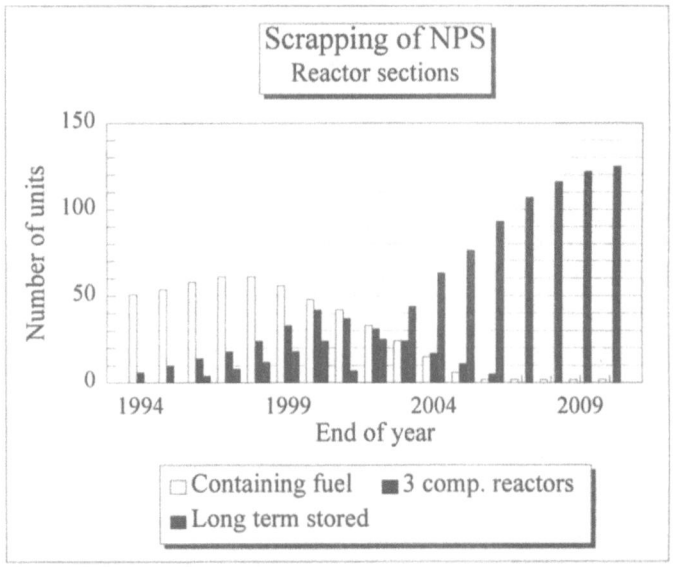

Figure 3. Disposal Rate of Russian Nuclear-Powered Submarines (NPS)

The total towing operation necessary to bring the 90 submarines to the scrapping yard is, depending on the yards involved, between 11,000 and 18,000 nautical miles. The

---

[1] During a meeting of the Joint Norwegian-Russian Expert Group on Radioactive Contamination of the Northern Areas in Severodvinsk in October 1995, the delegations visited a site for laid-up submarines and could inspect, from the outside, a nuclear-powered submarine (NPS) that had been laid up for eight years. It appeared to be in reasonably good shape.

longest single tows are about 350 nautical miles. The total amount of towing is not more than can be accommodated by one tug.

Approximately three years after running the reactor, fission products in the fuel are sufficiently decayed that heat generation is reduced to a manageable level. Manageable in this context means so low that active cooling systems are no longer needed. Heat generation persists, but there is no longer a need for active cooling systems to keep fuel temperature at an acceptable level. An implication of this is that local, intermediate storage is not necessary when handling the fuel from decommissioned submarines, the majority of which have been out of service for more than three years. In principle, the SNF can be loaded directly into the transport containers.

The submarines presently decommissioned contain a total of 140 reactors or about 35,000 fuel assemblies. Since a nuclear reactor is designed to contain nuclear fuel, this SNF is, in principle, stored in a safe place. However, the reactor, as well as the submarine itself, needs maintenance to remain safe and intact. The necessary control systems must work, for the initial period the fuel must be cooled, the submarine must be guarded to prevent accidents and acts of sabotage, and it must be properly maintained to prevent it from sinking. Some submarines have been laid up with SNF in their reactors for as long as around ten years.[1]

Assuming that the fuel assemblies are designed for a normal cycling time of, say, five years, and also assuming what one may call normal engineering practice with regard to design margins, there are strong reasons to believe that a situation is approaching where the design life for fuel elements is systematically exceeded. The characteristics of a fuel retrieval operation may change dramatically when that becomes the case. It may not be possible to carry it out in the planned way.

The Kværner/Energia group believes that corrosion of fuel assemblies and control rods to the extent that leaks of high-level wastes outside the reactor or recriticality of the SNF will be caused, will take longer than corrosion or breakdown of critical hull penetrations. Thus the Navy will face, and have to solve, the problems of sinking submarines before serious fuel-related problems start appearing. As the sinking of submarines still containing fuel obviously must be avoided, the Navy is in a situation where actions are absolutely necessary. Either the submarines must be defuelled, or an active maintenance program for the purpose of preventing their sinking must be initiated.

2.1. THE REACTOR SYSTEM

Because it is exposed to intense neutron radiation over an extended period of time, the reactor vessel itself and its internal parts also become radioactive. The induced radionuclides are *activation products*—which are different from the *fission products* in the SNF—because they are produced by interactions between neutrons that escape the reactor core and the construction materials of the reactor parts. The most important activation products generally have comparatively short half-lives of 100 years or less.

The radioactivity resulting from the decay of these activation products is significant. To ease the handling of the reactors, they are not dismounted, but rather remain in their location inside the pressure-hull. The hull is cut circumferentially, resulting in a reactor section (RS) that typically weighs 1200 tons.

The Russian plan is to place all contaminated parts that were removed from the reactor during cutting and cleaning operations inside the RS. Then, it will be sealed off by bulkheads at each end. Current plans recommend storage of these RSs out of the water, in a tunnel at the Kola peninsula. This tunnel can accommodate the total number of reactor sections.

2.2. LAND-BASED STORAGE FACILITIES

The Northern Fleet has one land-based SNF storage facility at Andreev Bay in the Zapadnaya Litsa Fjord. This facility has been described as "morally and technically obsolete" and filled to capacity. It reportedly contains about 70 reactor cores (equivalent to almost 20,000 SNF assemblies).[2] The fuel assemblies are stored in stainless-steel containers in a large stainless-steel clad concrete tank. A recently released report provides a description of the rather unsatisfactory conditions at this facility.[2]

Russian plans advise emptying the Andreev Bay storage facility and constructing a new one.

3. Handling of Spent Nuclear Fuel

3.1. DEFUELLING OPERATION

Obviously, there are different ways of looking at the procedures for handling SNF. One, irrelevant to Kværner/Energia's work, reflects the needs of an operative fleet of nuclear submarines. The following text only deals with operations relevant for the decommissioned submarines.

The removal of the SNF is one of the first steps in the process of disposing of a submarine. As stated above and confirmed by the Russian side, this requires "complete cooling" of the nuclear power plant. In practice, this probably means that no reactor is defuelled until at least three years after it was last run. At that time, the short-lived activity in the fuel has decayed to a level where active cooling of the SNF is no longer required. This simplifies the subsequent handling of the SNF.

Plans for the disposal program can assure that nearly all SNF is more than three years old when a nuclear-powered submarine (NPS) is defuelled.

---

[2] Figure reported by the Russian side during the visit of Kværner Moss Technology a.s. to Moscow and Severodvinsk in October 1995.

The technical details of the defuelling process are not presented by the Russian side, but the general principles are discussed. Defuelling takes place while the submarine is still afloat at the pier. A special vessel (Engineering Depot Ship, or EDS) is positioned on the outward side of the submarine. This EDS is the base for all defuelling work. The fuel assemblies are lifted out one at a time. On board the EDS, they are placed inside canisters which are again loaded into the TK-18 transport container. The EDS is equipped to accommodate one TK-18 at a time. This limited capacity for handling the TK-18 container is one of the major bottlenecks in the overall scheme. It necessitates the introduction of a dedicated TK-18 carrier.

The handling of the SNF includes two transfers, first from the reactor to the rack system in the EDS, then to the transport container (TK-18). The lifting of SNF is a risky operation, and the number of lifts should be minimized.

Participants from the Russian side have discussed a method whereby the fuel assemblies are loaded more directly into the TK-18. The same method is probably considered by the Russian Ministry of Atomic Energy (MINATOM), as reported in [4]. In addition, the proposed Russian design of a barge for TK-18 transportation includes features enabling it to take part in such an operation.

3.2. TRANSPORT CAPACITIES

One special train usually consists of four wagons, carrying a total of twelve TK-18 transport containers. This is sufficient for two reactor cores, equivalent to the SNF from one submarine.

According to available information, the transport container is built to standards equivalent to the best Western practice. In total, more than 50 are available.[3] A first estimate indicates that 24 containers will be continuously occupied by the handling of SNF from the decommissioned submarines. One set of 12 will be on the EDS, being loaded with fuel, while the other set of 12 is on the train or in Mayak being emptied.

One round trip to Mayak takes approximately 25 days. As the train unloading/loading operation takes five days, the transport part of the SNF-handling chain has an overall capacity of about one submarine per month. However, the defuelling operation is stated to take about 35 days and is, thus, the rate-determining step. An initial assessment of overall capacity is in the range of ten submarines per year.

The proposed SNF carrier will be built to an adequate ice class and will, therefore, also be able to operate in winter. The defuelling itself is a delicate operation which probably cannot take place under adverse weather conditions. For this reason, a detailed understanding of the overall capacity of the chain has not yet been established. However, there appears to be a good balance between the defuelling operation and the transport

---

[3] Reported by the Russian side during a meeting in Moscow, October 1995.

operation, and the total capacity appears to be sufficient for the overall goals to be achieved, disregarding the emptying of the Andreev Bay storage facility.

It has been stated that the emptying of this storage can take place at the same rate as the defuelling of submarines. In that case, the overall balance between the fuel handling operation and the transport operation can be maintained. The total amount of fuel will increase, and the overall goal of completing fuel operations by 2010 probably cannot be met. Furthermore, the methods for emptying the storage must be investigated before any qualified assessments can be made. The Andreev Bay operation is not routine.

This small capacity for transportation is partly a result of the recent introduction of the new container, TK-18, which cannot be loaded onto the wagons used for the older TK-12 and TK-13 containers. The old containers are no longer approved for use by the Russian authorities.

3.3. THE MAYAK REPROCESSING FACILITY

As described above, Russian intentions are to send all submarine SNF to be reprocessed at the Mayak Chemical Combine (also known as Chelyabinsk-65) near Ozersk in the southern Ural Mountains.

As also mentioned earlier, in Russia, SNF is considered to be a *resource* rather than a waste product. During the reprocessing of uranium, neptunium and plutonium are extracted from the SNF. The extracted uranium, and possibly some of the plutonium, is used in the production of new nuclear fuel. The fission products in the SNF make up the waste from the reprocessing, yielding highly radioactive liquid waste.

Transport from northwestern Russia to Mayak is, at present, a significant problem. The reprocessing facility includes three lines which can process SNF from commercial power reactors as well as special fuel from naval, research and highly-enriched uranium reactors. Currently, the capacity for reprocessing heavy metal from standard uranium-aluminum naval fuel has been stated in figures ranging from eight to fifteen tons per year, depending on the source of information. (See [3] for an example.) This is in the range of being sufficient for the overall goals to be achieved. However, no detailed knowledge is available regarding the amounts of fuel to be received from the Pacific Fleet and from submarines in operation. It is known, though, that the Russian side wishes to establish an intermediate storage facility for submarine fuel at Mayak and has declared that such storage is necessary in order to be able to receive fuel from decommissioned submarines at the anticipated rate.

An ongoing problem, about which ambiguous information has been provided, is Mayak's capability to process all types of fuel. Zirconium-based fuel and fuel from metal-cooled reactors may not be able to be processed. Also, damaged fuel can obviously not be processed. This establishes the need for some SNF storage of an intermediate or more long-term nature. Work is currently underway to solve the problem of the zirconium-uranium alloys.[3] There appears to be no available information as to how large a fraction of the fuel can actually be reprocessed by current methods.

As to the burden on the environment due to present operations of the Mayak facility, this is a complex question which must furthermore be weighed against the environmental effects resulting from intermediate storage and eventual deposit of the SNF in a repository. The Kværner/Energia group considers it to be outside of its competence to judge this difficult question.

4. Present Status of Disposal Accomplishment

4.1. ORGANIZATION AND RESPONSIBILITIES

A clear definition of responsibilities cannot be deduced from the information received during this project. A summary of the joint Kværner/Energia understanding, however, is given below. The overruling principles are:

The Navy assumes responsibility for the fuel from the moment it arrives at the Navy's storage facility until it is shipped for reprocessing. Of relevance to the current study is that the Navy is responsible for the safety, security, and quality of the following operations:

- Defuelling of decommissioned submarines;
- Interim storage of spent fuel;
- Loading of SNF into transport containers and transporting them to the railway terminal; and
- Transport and storage of reactor sections;

MINOBORONPROM (the Ministry for the Defense Industries) operates all major shipyards and is responsible for the scrapping work. This will include intermediate storage and treatment of low-level wastes generated during the scrapping operation. The Ministry's research institutes and design bureaus are responsible for the integration of reactor systems and fuel management with the technologies and operations of naval vessels. As MINOBORONPROM is the body executing the submarine scrapping operation, it can be inferred that MINOBORONPROM or its institutes bear responsibility for solving any situation or condition arising from this disposal which cannot be remedied by normal procedures or institutions.

The Ministry of Atomic Energy (MINATOM) is responsible for the nuclear fuel in general, i.e. after receiving it at the railway terminal. MINATOM also issues the regulatory documents governing the handling of such material.

4.2. PRESENT PRACTICE

Decommissioned NPSs are, at present, treated in one of two different ways:

- Either they are prepared for prolonged floating storage, without cutting out the reactor compartment; or

- The reactor section is cut out and prepared for floating, intermediate storage. The other parts of the hull, fore and aft of the reactor section, are scrapped.

By the beginning of 1995:

- Nine submarines had been prepared for prolonged floating storage;

- Six submarines had had the reactor sections cut out and were under intermediate, floating storage; and

- The hulls of three submarines were under scrapping at Zvyozdochka and one at Nerpa.

4.3. PRESENT PRACTICE, REMOVAL OF SNF

As changing of reactor fuel is a well-defined, routine operation, the final defuelling of a decommissioned submarine is not anticipated to require new procedures or technology.

4.4. INTERIM STORAGE OF SPENT NUCLEAR FUEL

The Russian side has, on some occasions, presented the view that there is a need for an intermediate storage facility for SNF in the Kola region. As Kværner Maritime a.s. is occupied in these matters for the sole purpose of assisting in the prevention of future environmental problems, it approaches alternative strategies from this perspective.

From the Kværner point of view, four alternatives exist:

1. Transport of all SNF to Mayak for immediate reprocessing at the rates allowed by the transport capacity and reprocessing capacity.

2. Transport of all SNF to Mayak at the rate allowed by the transport system. Reprocessing capacity may not be adequate, and a buffer storage site may be required at Mayak.

3. As above, but the buffer storage site is located in the Kola region. This alternative will result in a reduced need for transport capacity out of the region.

4. Most, or all, SNF is stored for a prolonged time. The storing is not motivated by reprocessing capacity, but by a wish to delay the operation, thereby enabling Russia to strengthen its economy or to improve general conditions at the Mayak plant. It should be noted that prolonged storage under safe conditions will facilitate all subsequent handling. Even if this places a burden on future generations, the overall burden could be smaller than what arises out of immediate reprocessing under conditions that are not optimum.

5. Conclusions

In general, the Russian plan for disposing of the decommissioned submarines does not contain elements with a high level of risk or which can reasonably be described as not feasible. The Kværner/Energia evaluation is that it is not improbable that the proposed elements can be built and operated as described in the plan, achieving the planned disposal capacity.

As reported in [1], the United States Navy apparently performs a very efficient scrapping procedure. Using one specialized dry-dock, it removes the reactor section, recovers usable equipment, and cuts the hull into ten- to fifty-ton pieces for eight submarines per year. Resource allocation is not quantified, but it probably does not exceed 35,000 person-hours per submarine. It is stated that the scrapping costs are US$6.9 million per nuclear submarine.

In the Kværner/Energia context, a direct comparison with methods and technology used in the United States is not relevant. The industrial environments are very different. However, the figures give an indication of attainable resource consumption in a well-organized scrapping process.

A concept based on the upgrading of one yard to be the leading yard and on the use of the TK-18 container for all transport purposes has emerged during the cooperative phase of the project. It is the practical implementation of this scenario which forms the recommendations from Kværner/Energia's work.

Based on the above, the following will be important contributions to safe handling of the SNF:

• Construction of a special container vessel for transportation of spent fuel in TK-18 containers; and

• Construction of a few special train-wagons for transportation of spent fuel in same.

The proposed units can also be utilized, serving the same important functions, in a future emptying of the Andreev Bay storage site. This item is only indirectly related to the problems of the decommissioned submarines, but is equally important from an environmental point of view.

The Kværner/Energia group would also like to emphasize the importance of developing methods for retrieving and handling fuel that, for various reasons, are no longer standard. The personnel involved in these clean-up operations will be facing potentially very difficult tasks.

The two projects above, plus the ones listed in the introduction, are now being further developed by Kværner/Energia in order to ensure an efficient implementation once the agreement between Norway and Russia is signed.

References

1. MacKinnon, M. III and Burritt, James G. (June 1995) Overview of Nuclear Submarine Inactivation and Scrapping/Recycling in the United States. NATO Advanced Workshop, Moscow.

2. Nilsen, T., Kudrik, I., and Nikitin, A. (1995) *Zapadnaya Litsa*, Working Paper No. 5, The Bellona Foundation. Oslo, Norway.

3. Office of Technology Assessment. (1995) *Nuclear Wastes in the Arctic: An Analysis of Arctic and Other Regional Impacts from Soviet Nuclear Contamination*, Report No. OTA+ENV+623, , Congress of the United States, U.S. Government Printing Office, Washington, DC, USA, Chapter 4.

4. Yegorov, N.N. (1995) *Plenary Address to Seminar on Nuclear Waste Management in the Russian Federation*. Vienna.

CHAPTER 4
THE *LEPSE* PROJECT

Result of the European Commission Study for Retrieval of Spent Fuel
TACIS Program

HENRI DE LA BASSETIERE
International Nuclear Division, SGN Reseau Eurisys
St. Quentin Yvelines, Cedex, France

1. Introduction

The *Lepse* is a service vessel in the fleet operated by the Murmansk Shipping Company which is currently used to store spent nuclear fuel from icebreakers. Construction of the ship began in 1934 but was not completed, and the ship was sunk in 1941. It was salvaged after the war and modified for nuclear service purposes, and then brought into use in 1962. In 1967, fuel elements from the *Lenin*, which had been damaged during a loss-of-coolant accident, were transferred to the *Lepse*. These elements had swollen and become distorted during the accident, and it proved necessary to use some force to insert them into the storage channels provided on the ship.

The ship now contains some 645 nuclear assemblies, which generate about four kilowatts (kW) of heat and have a radioactive content of some 30 PBq (750,000 Ci). It also has some storage areas for liquid and solid radioactive wastes, which are estimated to contain a further 0.8 TBq (20 Ci) of activity. The condition of the ship is a matter of some concern. There has been some corrosion damage to the hull. It is located adjacent to a busy shipping lane. A collision or fire onboard resulting in the flooding of the ship could lead to its capsize.

For these and other reasons, it has been regarded as a matter of first priority to remove the spent fuel from the *Lepse* and later to decontaminate and decommission the ship itself.

The *Lepse* is a vessel of length 82.9 meters and beam 17.1 meters, with a displacement of 5,600 tons (fully loaded). The spent fuel is contained in two tanks, each a cylinder of diameter 3.5 meters and height 3.4 meters, within which are installed 366 channels of diameter 76 millimeters and 4 "caissons" of diameter 485 millimeters. Shielding is provided by a surrounding metal box and by a revolving plate 280 millimeters thick, through which there are penetrations with removable plugs that give access to the stor-

E.J. Kirk (ed.), Decommissioned Submarines in the Russian Northwest, 41–47.

age channels and caissons. (The channels are arranged in a series of concentric rings, and each penetration gives access to all the channels in a given ring).

This system was designed to permit fuel elements to be introduced and removed by means of a shielded transfer flask lowered onto the penetration. Unfortunately, a significant fraction of the fuel elements (estimated at 70 percent) are now jammed into the channel and cannot readily be removed by the design procedure.

Conditions inside the tanks reflect the history of operations since the fuel was introduced. Initially, the tanks were flooded, and cooling was provided by two circulation loops of water. In 1990, however, a leakage problem developed in the primary loop and the channels were drained, so the fuel is now relatively dry. Corrosion within the tank has led to deposits of corrosion products on top of the fuel and may have led to deterioration of the condition of some of the fuel elements themselves.

2. The Study Financed by the European Union Under The TACIS Program

The project WW.93.06/01.05/B003 was established within the TACIS program to undertake a preliminary study of the feasibility of:

- removing the spent fuel;
- transferring the fuel to Mayak for reprocessing; and
- decommissioning the *Lepse*.

The contract was awarded to SGN, as project manager, and AEA Technology in October 1995, with a completion date of July 1996. The Murmansk Shipping Company has been closely involved in the work and has provided much of the relevant operational data. The project is now about 80 percent complete, and the main outlines of the proposed solution to the first objective—removal of the spent fuel—are now clear. Since removing the fuel will eliminate at least 99 percent of the radioactive hazard, it seems reasonable to start making concrete plans for the implementation of this part of the work now.

3. Removal of the Spent Fuel

Three alternative approaches have been evaluated.

One scenario was rejected at an early stage—to grout the entire contents of each tank as they stand, then transfer the resulting concrete block to a suitable long-term repository. This option would require the section of the *Lepse* hull to be removed and would therefore require the construction of a dry-dock facility. This option was judged to lead to an object for disposal which would be unacceptable to the Regulator of any long-term repository.

A second scenario was to install a temporary modular containment system above the tanks, within which the following operations would be undertaken by remote handling (telerobotic) means :

- installing a gantry manipulator capable of accessing all the fuel channels;
- removing the steel cover plate;
- cutting free each channel;
- drying the removed fuel assembly;
- withdrawing each channel into a shielded transfer flask; and
- removing the flask to canistering and a handling facility, within which the fuel could be transferred to a transport container.

This approach was judged to be technically feasible, but the problems of ensuring that adequate radiation shielding was provided to protect the operators and the general public, without compromising the stability of the *Lepse,* were found to make it more expensive, both financially and in terms of dose uptake by operators. The cost was found to be 25 percent higher than that of the preferred approach.

The recommended scenario is the third—to leave the steel cover plate in position and use the existing penetrations to access the fuel.

The proposed sequence of events is as follows:

1. Clear the area above the steel cover plate and install an improved air ventilation system.

2. Install some additional shielding on top of the steel cover plate to reduce the operator dose to an acceptable level.

3. At each fuel channel position, clear away the debris covering the fuel element using a shielded tool inserted through the existing penetrations.

4. Lower into position a shielded mechanical cutting device that contains a tool capable of cutting the weld which holds each channel in position within the tank.

5. Lower into position a shielded transfer flask containing a handling tool capable of gripping and then withdrawing the channel including the enclosed fuel assembly.

6. Transfer the channel to an adjacent vacated caisson which can receive a canister and in which a drying system has been installed that is capable of dehydrating the fuel and channel to the extent demanded by Mayak.

7. Simultaneously dry three fuel assemblies directly in the canister and complete the canistering operations.

8.   Transfer the canister within a shielded transfer flask to a TK-18 packing station.

4. Transporting Fuel to Mayak and Decommissioning the *Lepse*

These two topics are within the part of the scope of the study contract which remains to be addressed in detail, though it is already clear that they do not pose any insuperable problem. The Murmansk Shipping Company (MSCO) has already agreed with the management of Mayak that the fuel is, in principle, acceptable for reprocessing, notwithstanding its damaged condition.

The decontamination of the ship prior to its decommissioning is not an urgent problem and will require planning and preparation because of remaining radiation levels in some areas, but once this has been completed, it should be possible to break up the ship by employing conventional marine engineering procedures.

5. Implementation of the Project

The study contract described above has addressed the problem in sufficient detail to both demonstrate a feasible engineered solution and to permit the scope of work of the implementation phase to be clearly defined. It is envisaged that the implementation project will consist of the following stages:

1.   Project Management and coordination between the Western and Russian companies involved.

2.   Detailed engineering studies, leading to documentation which will:

     * permit the purchase and/or manufacture of equipment;
     * enable MSCO to obtain authorization from Russian safety authorities; and
     * define an operational plan for each stage of the work, including the logistics of any transport arrangements.

3.   Support of MSCO to obtain authorization from Russian safety authorities.

4.   Inactive mock-up/commissioning trials of all safety-critical steps in the project, in order to:

     * define or confirm certain parameters of the operations, beyond what can be established by conceptual studies (e.g. the forces required to remove channels from the tanks); and
     * demonstrate the safety of the proposed procedures to regulatory authorities.

5. Procurement in Western countries and Russia and follow-up of fabrication under the Quality Assurance Program.

6. Elaboration of operation procedures.

7. Training of the MSCO operators to ensure that operators are familiar with all procedures before starting active work.

8. Installation and commissioning of equipment on board the *Lepse*, which will include:

   - preparation of the *Lepse;*
   - equipment to be installed on the *Lepse* (refer to the list of equipment); and
   - refurbishing of existing equipment to be used during operations.

9. The fuel removal operation:

   - technical supervision of the operations.

6. Proposed Timetable

The entire defuelling project is estimated to take about 3.5 years. However, this estimate is subject to a number of uncertainties:

- Time will be needed to put in place the various components of the funding which may be required.

- The time which will be required to obtain Regulatory approvals from the Russian authorities remains somewhat uncertain.

- The defuelling operations will be undertaken by MSCO staff who have parallel responsibilities for the continuing operations of the icebreaker fleet, and these two activities will have to be coordinated.

This time estimate of 3.5 years includes a certain contingency provision to cover these considerations.

7. Cost Estimates

The scope of work defined above has been costed based on the following assumptions:

1. The project management of the overall project activities will be undertaken by a single European Union (EU) contractor.

2. All site operations will be undertaken by MSCO staff with EU technical assistance.

3. All necessary Regulatory approvals will be obtained by MSCO, using technical documentation and support supplied by the EU contractor.

4. All transport casks for sealed canisters and lifting gear will be supplied by MSCO.

5. All other equipment will be procured in Russia and Western countries by the EU contractor.

6. There will be no unforeseen delays in obtaining authorizations or in the availability of MSCO staff.

7. There will be no change in the information currently available to the contractors at this intermediate stage in the study contract.

On this basis, the cost estimates based on European Standards can be broken down as follows:

Identified equipment	3770
Provision for other equipment (e.g. backup, upgrading, handling system)	1000
Canisters	1700
Engineering	1815
Technical assistance during operations	430
Total cost	8715 MECU

8. Contractual and Project Management Aspects

It seems likely that to obtain funding on this scale will require collaboration between a number of different interested parties. This raises the question of how these parties will wish to coordinate their contractual arrangements. In meetings with possible funding agencies to date, the following have been discussed, on principle:

*   need for an agreement between the funding agencies;
*   appointment of an engineering company responsible for the whole project, with the power to place subcontracts as necessary within the overall funding resources allocated to it; and
*   dividing the project into a number of "independent" subprojects, each of which would be the responsibility of a single funding agency.

There will need to be an effective project management system for the whole project, since all technical aspects of the project will be interrelated, whatever the funding and contractual arrangements may be. In deciding, it will be desirable to bear in mind the time which will be required to reach the necessary decisions in each case, and to implement the associated contractual arrangements.

9. Conclusions and Recommendations

It is concluded that the work which has already been completed on the EU-funded study on the *Lepse* is sufficiently advanced to permit the development of plans for implementing the most urgent part of the work—removal of the spent fuel.

It is estimated that this work will require about 3.5 years and will cost about 8715 MECU, based on the assumptions set out in this paper. The Advisory Committee for the *Lepse* project is therefore invited :

1. To endorse the plans set out here for defuelling the *Lepse;* and

2. To consider how funding could be made available on the timescale required to ensure a prompt reduction in the level of nuclear hazard which the *Lepse* currently represents.

CHAPTER 5
INTERIM STORAGE OF SPENT NUCLEAR FUEL IN THE ARCTIC FAR NORTH

A.P. HOSKINS, R.A. RANKIN, B.G. MOTES, J.O. CARLSON
Lockheed Martin Idaho Technologies Company
Idaho Falls, Idaho, United States
C.W. LAGLE
Lockheed Martin Environmental Systems & Technologies Company
Houston, Texas, United States
K.R. JOHNSEN
Kværner Maritime a.s.
Lysaker, Norway

1. Introduction

A United States Congress Office of Technology Assessment (OTA) report [1], issued in September 1995, states that: "All major nuclear nations face nuclear waste problems. Many also share a common history of radioactive contamination incidents stemming from inadequate attention to environmental protection. The United States and Russia, in particular, have some similar nuclear waste management and contamination problems..."

When discussions of potential and real contamination problems in the Arctic Far North occur, the "Arctic" elicits images of vast frozen expanses with little human habitation or industry and a relatively pristine environment. These images are seldom accurate, and contamination from both military and industrial activities has brought questions about its impact not only locally but in the wider Arctic region.

Whereas past dumping of nuclear materials in Arctic areas has received considerable attention recently from scientists and analysts, the risk of future releases has not been subject to scrutiny or careful study.

Because it is so difficult and costly to take useful actions after radionuclides have been released into the environment, it is wise to consider prevention efforts now that could minimize future accidents, releases, or discharges. This paper focuses on the fuel storage aspects associated with the prevention of future releases of radioactive material into the environment.

E.J. Kirk (ed.), Decommissioned Submarines in the Russian Northwest, 49–55.
© *1997 Kluwer Academic Publishers.*

2. Existing Concerns in the International Community

Against this backdrop, the United States and the international community are directing attention and resources toward problems arising from nuclear activities in Arctic regions. The OTA has selected three key areas that appear to be the most significant to future contamination risks at this time:

1. the Russian Northern and Pacific Nuclear Fleets with regard to their vulnerabilities to accidents during the downsizing and dismantlement activities now underway;

2. the management of spent nuclear fuel and the waste from these fleets, including concerns about effective containment, safety, and security, and future releases; and

3. the possibility of accidents or releases from Russian civilian nuclear power plants, particularly those located in the Arctic.

During the past three decades, the Soviet Union built the largest fleet of nuclear submarines and the only fleet of nuclear-powered icebreakers in the world. The Russian Navy has been retiring and decommissioning older nuclear submarines at an increasing rate over the past several years. More than 120 Russian nuclear submarines have been taken out of service, and many are in various stages of dismantlement. Only about 40 of these have had their spent nuclear fuel removed. Some of these submarines have been out of service with nuclear fuel aboard for more than 15 years, primarily due to lack of storage facilities and equipment needed to dismantle and transport the fuel.

Removing this fuel from reactors for temporary storage and selecting or developing appropriate future treatment or storage technologies are challenging and costly procedures that will require some technology not available in Russia. Investigations of the situation at local bases of the Russian nuclear fleet in the North and Far East show that several problems exist which, if not addressed soon, could lead to accidents or pressure to engage in more dumping of waste.

With these driving considerations, the defense ministries in Norway, Russia, and the United States have agreed to cooperate in dealing with radioactive contamination from military sources in the northern seas. This specifically involves work on the Kola peninsula in northwestern Russia, bordering northern Norway. In the Kola region, existing storage facilities are virtually filled to capacity. One of the main tasks facing the Russians is to build waste storage and disposal facilities for all types of radioactive waste from civilian and military sources.

The Northern Fleet of the Russian Navy has many nuclear-powered vessels (most of which are submarines) that have been taken out of service. This figure will rise to approximately 100 during the next few years. Due to the lack of storage capacity for radioactive waste, the vessels have not been decommissioned, but remain at their berths, in most cases still with high-level spent uranium fuel in their reactors.

With the continuation and expansion of international efforts to address spent fuel problems in the Russian North (i.e. the Kola peninsula and Murmansk), some significant improvements are possible in the prevention of future radioactive releases there.

With that goal in mind, Norwegian, Russian, and United States companies are considering projects that will reduce overall environmental risks.

3. Companies Involved

3.1. KVÆRNER MARITIME

Kværner Maritime is a Norwegian technology company within the Kværner Group with special knowledge about containment systems for hazardous cargos, maritime operations, design of ships, and floating structures. Kværner Maritime has a number of projects with Russian companies with Russian interests. Together with the Russian company Energia, Kværner Maritime has carried out a study for the Norwegian Ministry of Foreign Affairs on the disposal of Russian nuclear submarines. This work was carried out in close cooperation with Russian authorities.

3.2. LOCKHEED MARTIN IDAHO TECHNOLOGIES COMPANY

Lockheed Martin Idaho Technologies Company (LMITCO) operates the Idaho National Engineering Laboratory (INEL) for the United States Department of Energy (DOE). The INEL is located in the northwestern region of the United States and employs about 8,000 people, many of whom are scientists and engineers. Its mission during the past 40 years has focused on nuclear reactor and nuclear fuel cycle research, including spent nuclear fuel reprocessing and storage. During this time, 52 different nuclear reactors have been built and operated at the INEL. They were built for propulsion and power research, and none were used to produce weapons materials. One of LMITCO's current missions is to store and manage United States government-owned spent nuclear fuel (SNF). There are more than 80 different types of spent nuclear fuel at the INEL.

LMITCO has significant experience with the management of highly enriched spent nuclear fuel, primarily from the United States naval nuclear propulsion program. The INEL has stored and reprocessed this type of fuel for over 40 years. LMITCO personnel have hands-on experience with a wide diversity of spent nuclear fuel types and fuel matrices. These skills have been applied to fuel in a variety of conditions, from new fuel elements to severely corroded and damaged material from the March 1979 Three Mile Island Reactor 2 (TMI-2) accident.

LMITCO is dealing with many spent nuclear fuel challenges and issues. Each of the fuel types at the INEL has differing characteristics. Some of the fuel types are similar to those facing the Russian Federation. Most of the fuels at INEL are in good condition, but a small percentage of them are old and have been in storage for up to 30 years in less than ideal conditions. Some of these fuels came from reactors that experienced operational problems, such as TMI-2, where the fuel was severely damaged. Some fuels were intentionally disrupted during reactor fuel testing programs.

Some of the spent nuclear fuel has been in inadequate storage environments that have resulted in the corrosion of both the cladding material and, in some cases, the fuel meat. In a few cases, the spent nuclear fuel was stored in burial grounds where its storage conditions were not monitored and the SNF was not controlled to protect the environment.

As a result, LMITCO is proceeding with a program to recover and stabilize damaged or degraded fuel and place it into safe storage pending final disposition.

The INEL has demonstrated success in dealing with deteriorated fuel and the older spent nuclear fuel storage facilities. It has developed inspection methods using remotely operated micro-camera television and still photography as well as non-destructive methods (e.g. ultrasonics and eddy current) to examine spent nuclear fuel and containers. It has also developed methods and tools for retrieving—and has remotely retrieved—deteriorated SNF. The INEL shared the technology lead with General Public Utilities (the owner of TMI-2) in retrieving the TMI-2 fuel after the accident, shipping it to the INEL, and placing the damaged fuel into safe storage.

Other applicable INEL expertise includes:

- extensive experience with the recovery of damaged fuel from both inside the reactor vessels and in storage locations;

- experience in the decontamination of metal and concrete surfaces;

- extensive experience in nuclear safety analysis (including criticality safety) and modeling; and

- experience with remote inspection and robotics and with site monitoring and assessment.

INEL expertise also includes special expertise in fuel examination/characterization, corrosion mitigation, and recovery/treatment of damaged fuel.

Lessons learned from extended storage of DOE-owned spent nuclear fuel indicate that continued storage in a water environment is not acceptable. Many of the fuels and fuel containers have degraded in this environment. DOE, the United States commercial industry, and international experience, as well as technical data, confirm that transfer of these fuels to a dry inert gas environment is necessary and prudent for continued safe and cost-effective interim storage.

4. Proposed System

To handle fuel storage requirements through the next century, LMITCO is presently developing a modular dry storage system for a wide variety of fuel types to be stored on site. The interim dry storage system will effectively maintain confinement of radioactive material under normal, off-normal, and credible accident conditions. The system will include provisions to protect against nuclear criticality under the worst case moderator and reflector conditions. The storage system will maintain fuel and cladding below maximum allowable temperatures, which are dependent on the age of the fuel.

The INEL project is based on the use of dry storage canisters, which are placed inside storage modules and set on an outdoor pad for interim dry storage, ready for shipment when an off-site destination is identified. The modular dry storage unit, consisting of the storage canisters, the storage modules, and the transportation cask system are standard, commercially available items. Similar systems are in use at more than 15 commercial reactor locations within the United States.

Key features of this dry storage system are that it:

- is modular;
- is above ground;
- is prefabricated;
- can be expanded as needed;
- is designed to facilitate transportation of both the fuel and the modular structure;
- is easily decontaminated/decommissioned;
- will be licensed by the United States Nuclear Regulatory Commission (NRC) for interim storage; and
- provides for the waste form to be packaged for off-site transportation (i.e. "road ready"), which results in a passive storage system requiring minimal operating and maintenance expense.

Necessary steps leading to the use of such a system are:

- characterization of the fuel and definition of dry storage criteria;
- design of the internal canister basket for the specific fuel type and storage configuration;
- securing an NRC license for the specific fuel type and storage configuration; and
- initiation of procurement and fabrication of the canisters.

4.1. CANISTER

The canister is the key to the system. It functions as one of the containments for the fuel and serves a "dual use" or "dual purpose," as it is used for both transportation and storage. The canister is a standard size and shape. Its external dimensions will match standard dry storage canisters for commercial fuels. It has shield plugs in both ends to provide worker radiation shielding during transfer of the canister from the cask to the storage module. The basket inside the canister is custom-designed to the specific type of fuel to be placed in it, which allows the system the flexibility to store almost any fuel. Each significant type of fuel will require a separate safety analysis prior to interim storage. The canister is seal-welded shut prior to shipment.

4.2. STORAGE MODULE

The storage module, a concrete structure consisting of prefabricated components (base and top) that can be delivered to the site for installation, holds the canister. It is designed to limit radiation site doses and is ventilated to provide for passive convective cooling of the fuel. The storage module is designed to withstand design basis accidents (earthquake, weather, fire, explosion, etc.). It rests on a concrete pad.

4.3. TRANSPORTATION SYSTEM

The storage system requires a system-specific transportation cask for transporting the canisters to and from the storage module. The canister is transferred between the shipping cask and the storage module via a hydraulic ram (part of the transportation system). The cask is dual licensed by the NRC for use as both a shipping and a storage cask.

Design life for system components are:

- 60 years for interim storage modules and storage pad; and
- 100 years for dry storage canisters.

5. Dry Transfer Cell

LMITCO is also developing a dry transfer cell which will provide the capability to make dry transfers of INEL spent nuclear fuel inventory and future receipts from shipping casks into the dry storage canisters. The cell will provide equipment for drying, inerting, sealing, leak checking, and sampling of shipping cask internal gases.

The dry transfer cell will:

- handle transfers from a maximum of 12 incoming casks per month;
- have remotely changeable components; and

- allow for local maintenance of contaminated equipment.

It will receive spent nuclear fuel shipments by truck, but receipt by railcar is also possible. The cell will be designed to easily adapt to yet-to-be-identified shipping casks whose size and weight are enveloped by known casks. Siting of the cell will be based on access to existing utilities, use of existing roadways, access to adequate security perimeters, and access to a rail spur.

The INEL project will focus on fuels that are generally well-characterized, have sound cladding, and are easily dried. Additional development work is required to establish criteria for fuels needing treatment prior to placement into dry storage. (Future projects for placing "treatment required" fuels in interim dry storage may include a spent nuclear fuel preparation facility and procurement of additional dry storage canisters and modules).

6. Summary

The proposed system identified in Section 4 will be presented for consideration by Russian, Norwegian and United States authorities. It corresponds to priorities being pursued in a bilateral context between Norway and the Russian Federation. Such a cooperative effort will also correspond to the aim and priorities of the "Contact Expert Group on the International Radioactive Waste Projects" (CCEG), an activity under the auspices of the International Atomic Energy Agency (IAEA), for the promotion of radioactive waste management projects in Russia.

This type of spent nuclear fuel storage system—modular, prefabricated, transportable, and expandable as needed—could solve many of the challenges associated with lack of spent fuel storage in the Kola region. The Kværner Maritime/Lockheed Martin team is prepared to assist the Russian Federation in acquiring an interim dry storage system similar to the one presented in this paper.

References

1. (1995) *Nuclear Wastes in the Arctic: An Analysis of Arctic and Other Regional Impacts from Soviet Nuclear Contamination*, Report No. OTA+ENV+623, Congress of the United States, U.S. Government Printing Office, Washington, DC, USA.

2. (1995) Report to the *Storting*, Norwegian Ministry of Foreign Affairs Plan of Action No. 34. *Stortingsmelding nr. 34 (1993-94)*, a communication from the Ministry of Foreign Affairs to the Norwegian parliament detailing the Norwegian concerns with the hazards from nuclear activities in the northern region.

CHAPTER 6
MULTI-PURPOSE CASKS FOR POWER STATION FUEL

Possibly Also A Flexible, Economical Naval Fuel Management System?

NIGEL MOTE
NAC International
Norcross, Georgia, United States

1. Abstract

Nuclear electric power utilities in many countries have opted to store spent fuel on an interim basis pending the availability of direct disposal facilities or a change in the economic and/or political climate for reprocessing. Many are using, or have decided to use, dry storage systems. Multi-purpose casks, which are licensed for the storage and transport of spent fuel, provide a flexibility unmatched by any other fuel manage-ment system.

The multi-purpose cask technology developed for the management of power station fuel can be directly applied to the management of spent fuel from naval propulsion units. This takes advantage of the investment already made by commercial companies in completing the design and licensing requirements for utility applications. It thus represents the quickest and cheapest route to providing additional storage capacity for naval fuel. Storage capacity can be provided as needed and, when required, the spent fuel can be transported to other storage locations or processing facilities without any requirement for fuel handling operations. The casks can be reused as often as needed.

In 1993 and 1994, NAC used its NAC-LWT casks to ship spent fuel from Iraq to Chelyabinsk, Russia for reprocessing. In 1996, NAC shipped all of the spent fuel from the Georgia Institute of Technology Research Reactor in Atlanta out of the city prior to the start of the Olympic Games. Also in 1996, NAC shipped all of the spent fuel from the High Flux Test Reactor at Brookhaven National Laboratory to the Savannah River Site. All of these operations represent the application of casks originally designed for commercial power station fuel management to highly enriched fuel operations. Currently, NAC is in discussion with the United States Navy concerning the use of its Universal Multi-Purpose Container System (UMS) for United States naval fuel management.

E.J. Kirk (ed.), Decommissioned Submarines in the Russian Northwest, 57–68.

2. Introduction

Until the late 1970s, most commercial power plant operators outside the United States adopted a spent fuel management policy of immediate reprocessing and recycling of recovered products. In response to rising reprocessing prices, decreasing values of recovered products, concerns over proliferation risks, and a belief in the favorable economics of direct disposal, many utilities have since opted to store spent fuel on an interim basis pending the availability of direct disposal facilities or a change in the economic and/or political climate for reprocessing and recycling uranium and plutonium.

Spent fuel has traditionally been stored in water-filled pools located in the reactor building or fuel handling buildings, on reactor sites, or as part of large centralized facilities (e.g. Sellafield, La Hague, CLAB). Because the economics of pool storage are dependent on the size of the facility, the construction of additional separate pools on reactor sites has only been pursued in a few countries, such as Finland and Bulgaria. Most utilities providing additional on-site storage capacity for the reactors on the site have decided to use, or are already using, dry storage systems. There are several types of dry storage systems available. Each one has its merits depending on the requirements of the plant operator, which include the quantity and timing of spent fuel arisings, licensing requirements, specifications for local manufacture, and schedules for off-site shipment. Multi-purpose casks are licensed for both storage and transportation of spent fuel, thus offering a flexibility unmatched by any other fuel management system.

Metal casks have been used to transport spent fuel for over 40 years. They have an outstanding safety record and there exists extensive experience in their design, licensing, and manufacture. These casks are designed to maintain their integrity under accident conditions that may occur during fuel handling and transportation, while also meeting the necessary specifications for criticality control, cooling, and shielding. Figure 1 shows the main features of the 25-ton NAC-LWT cask used by NAC International for the transportation of commercial power plant and research reactor fuel.

Storage casks must guarantee long-term integrity during extended storage periods without allowing degradation of the fuel they contain. As they do not need to withstand transport accident conditions, they are not required to meet the structural standards for transport casks and are typically less expensive than transport casks. The disadvantage of such storage-only casks is that the fuel must be unloaded in a fuel handling facility prior to transportation from the site.

Casks designed for both storage and transport must combine the design features required for each of the individual types of cask and are typically more expensive than either. However, the combined capability removes the need for handling the spent fuel more than once, which allows the cask to be moved to another location with a minimum of effort, delay, and cost.

59

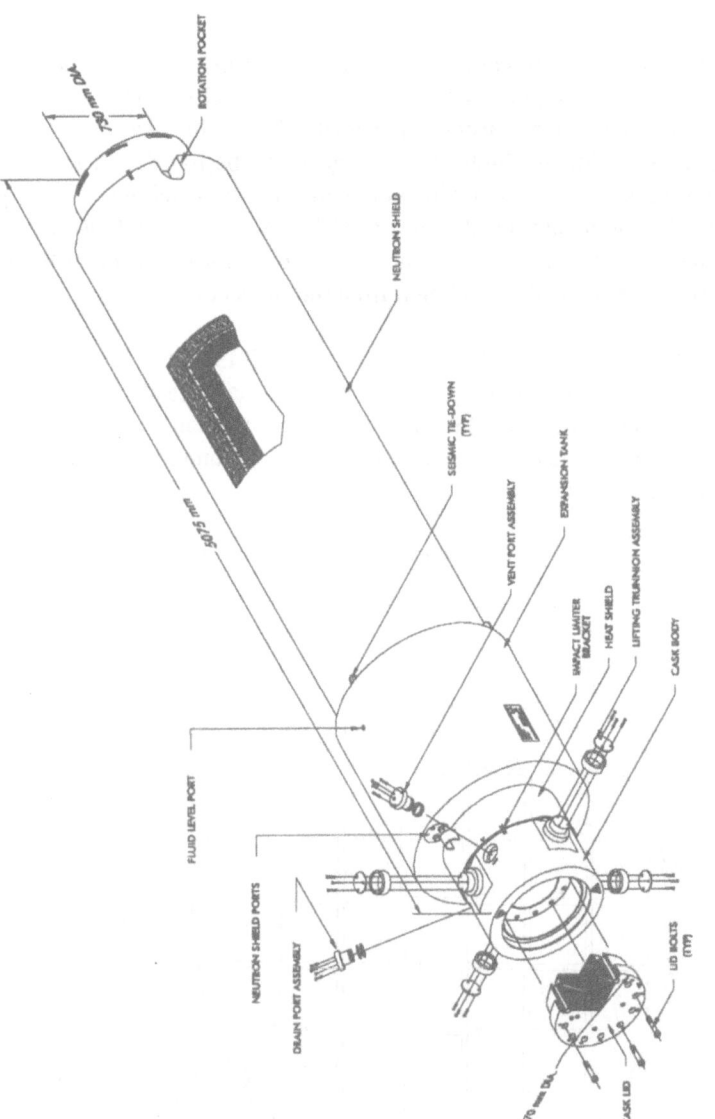

Figure 1. NAC-LWT Cask

60

3. Multi-Purpose Cask System

The most recent step taken in cask evolution is the development of the multi-purpose cask system. This system consists of a fuel basket housed in a seal-welded, but unshielded, canister. The filled canister can be loaded into a shielded concrete cask (also called an "overpack") for indefinite storage, and it can be transferred into a shielded metal transport cask (or overpack) without any further fuel handling. This is particularly attractive for the management of damaged fuel and fuel with failed cladding. It is possible that an additional overpack can be developed for disposing of the fuel in a permanent repository without unloading it from the canister.

NAC's Universal Multi-Purpose Canister System, or UMS, is the first such system to be developed. The UMS is unique in that its basket has already been approved for both storage and transport, while the transport overpack already has certificates of approval from both the United States Nuclear Regulatory Commission and the International Atomic Energy Agency.

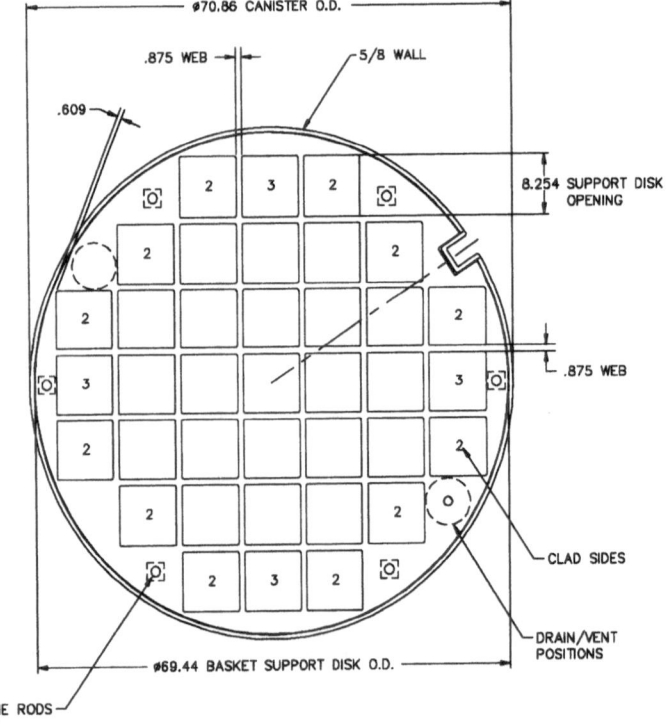

Figure 2. NAC-MPC (Yankee Rowe) Canister

Figure 2 shows a section of the canister designed by NAC to accommodate fuel discharged from the Yankee Rowe plant. Figure 3 shows the concrete overpack to be used for on-site storage pending determination of the fuel's ultimate destination. The

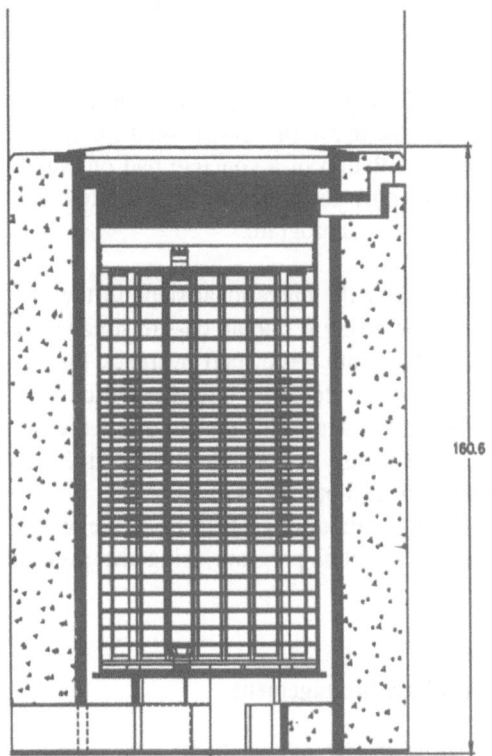

160.6

Figure 3. NAC-MPC (Yankee Rowe) Overpack

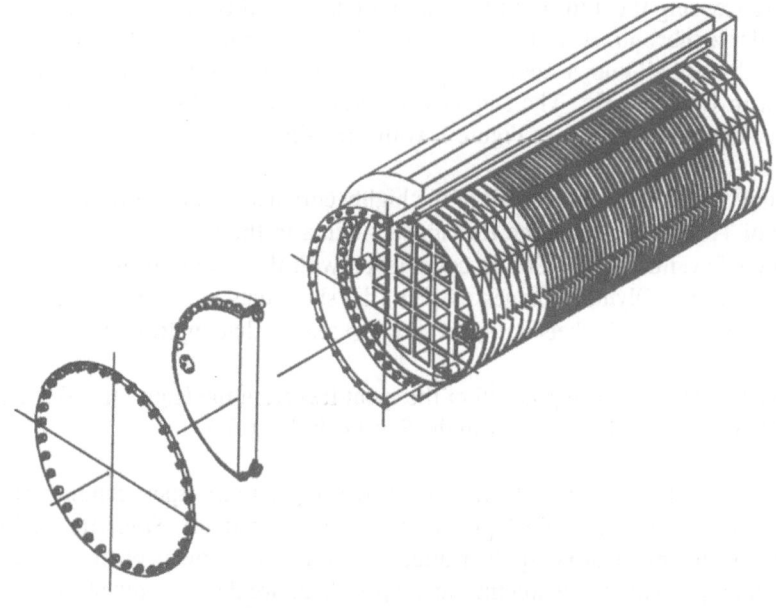

Figure 4. NAC Transportation Overpack

same canister can be used for transportation using the metal transport overpack shown in Figure 4.

With no need for fuel handling prior to shipment from the site, no operational equipment is required to support the use of this system on a storage site. The only construction requirement is the installation of a concrete pad on which to locate the storage casks. This means that the only decommissioning requirement is removal of the concrete pad at the end of the storage program.

Figures 5, 6, 7, 8, and 9 show the operational steps in the program for loading fuel into the UMS canister, transferring it into an on-site storage overpack and, after a period of storage, transferring it into a transport overpack for shipment to another site. If necessary, the fuel can then be stored for a further period before disposal, reprocessing, or even transfer to another site. This system offers the ultimate in flexibility for spent fuel management. The use of concrete overpacks for storage also makes the UMS inexpensive compared with other fuel management systems. Typically, only one metal transport overpack is required for each 15 or 20 fuel containers because containers are transported in a sequential program using the same transport overpack for many containers.

4. Application to Naval Fuel Management

All of NAC's casks were designed and originally licensed for commercial power station fuel. However, NAC has since extended the application of many of its casks, including relicensing them for use in shipping highly enriched fuel types. In 1993 and 1994, under a subcontract from the Ministry of Atomic Energy (MINATOM) of Russia, NAC shipped all known spent fuel from Iraq to Russia for reprocessing at the Mayak plant in Chelyabinsk. The shipment program used the NAC-LWT casks, which are shown diagrammatically in Figure 1, with specially designed and licensed baskets.

In February 1996, NAC shipped all of the highly enriched spent fuel from the Georgia Institute of Technology Research Reactor in Atlanta to the United States Department of Energy's Savannah River Site in conjunction with the requirements of the Atlanta Committee for the Olympic Games. Again, the shipments were completed using the NAC-LWT casks with baskets designed and licensed for that application.

Later in 1996, NAC also shipped all of the spent fuel from the High Flux Test Reactor at Brookhaven National Laboratory to the Savannah River Site.

These programs illustrate how the technology developed to transport commercial power station fuel has been used for highly enriched fuel operations. Similarly, multi-purpose cask technology can be applied directly to the management of highly enriched fuels, including naval propulsion unit fuel types. This would take advantage of the investment already made by the commercial industry and avoid the time and cost com-

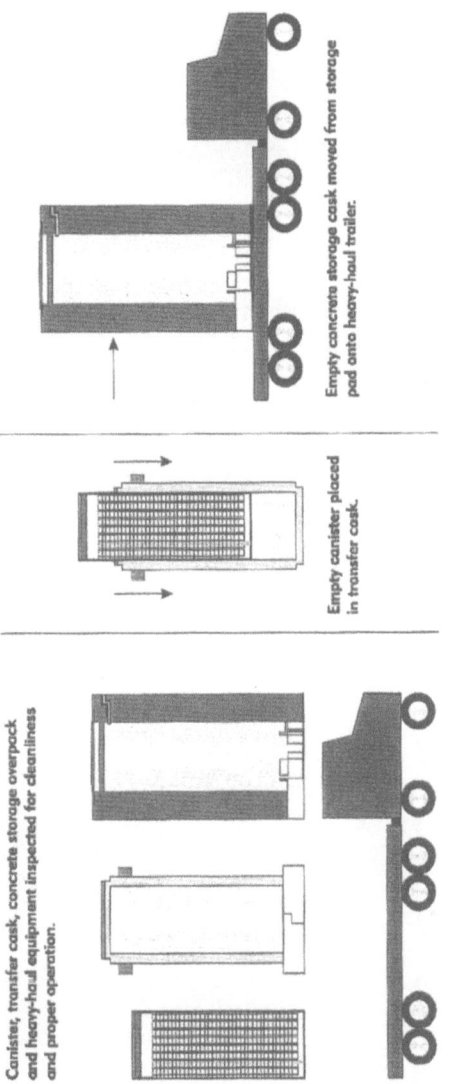

Figure 5. Equipment Inspection and Preparation

64

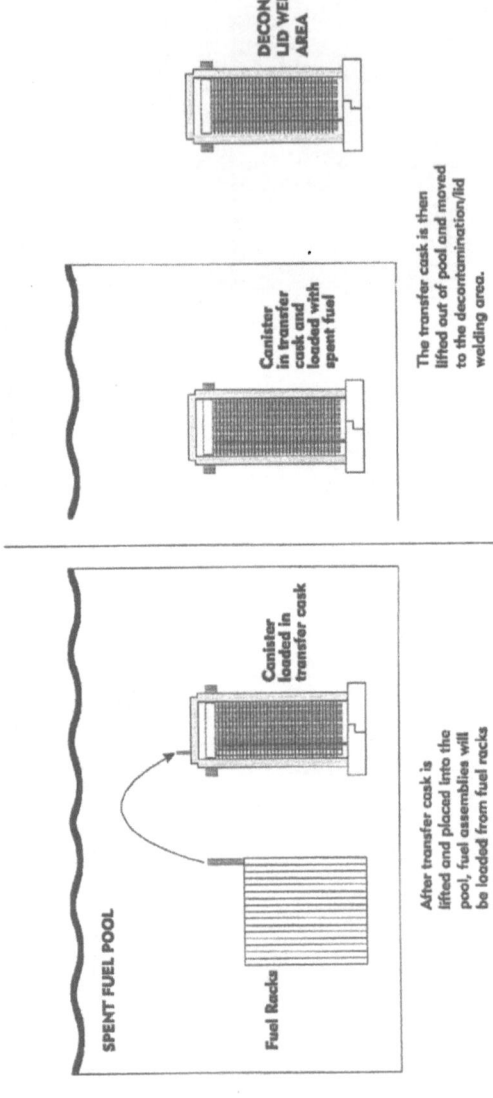

Figure 6. Fuel Loading

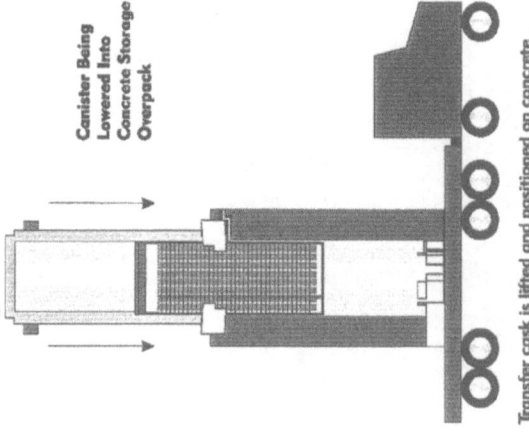

Canister Being
Lowered Into
Concrete Storage
Overpack

Transfer cask is lifted and positioned on concrete overpack. Transfer cask doors are opened, and the canister is lowered into the storage overpack. The transfer cask is removed, and the concrete overpack lid is positioned and secured.

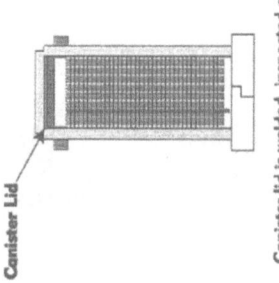

Canister Lid

Canister lid is welded, inspected and tested using a hydro test. Canister is flushed, vacuumed and backfilled with helium. Welds checked again using a helium leak test.

Figure 7. Canister Sealing and Transfer

66

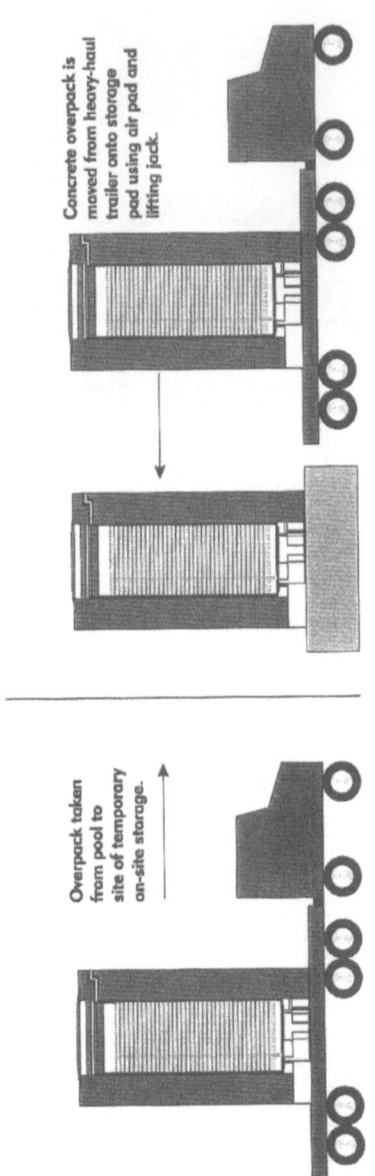

Concrete overpack is moved from heavy-haul trailer onto storage pad using air pad and lifting jack.

Overpack taken from pool to site of temporary on-site storage.

Figure 8. Cask Placement for On-Site Storage

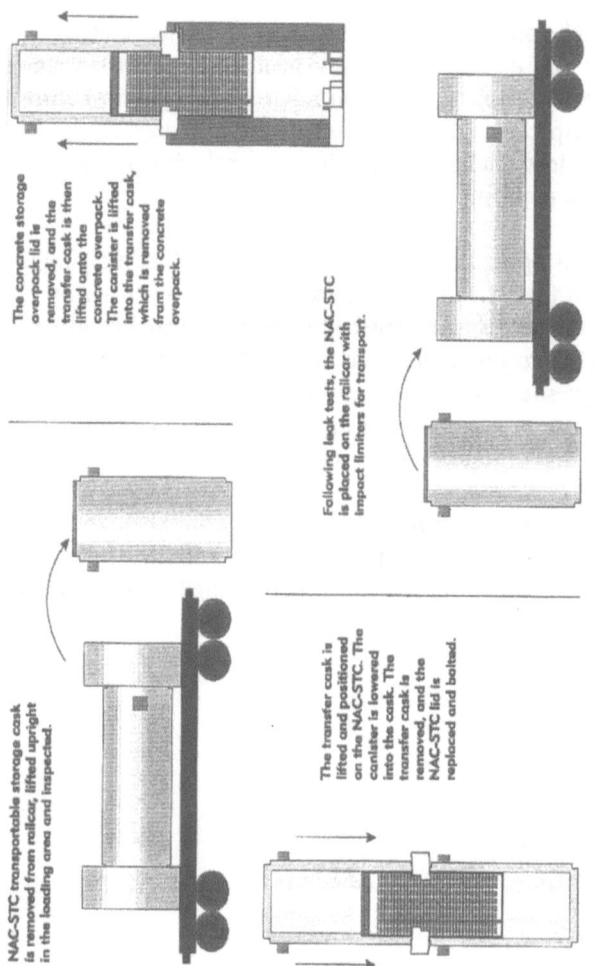

The concrete storage overpack lid is removed, and the transfer cask is then lifted onto the concrete overpack. The canister is lifted into the transfer cask, which is removed from the concrete overpack.

NAC-STC transportable storage cask is removed from railcar, lifted upright in the loading area and inspected.

Following leak tests, the NAC-STC is placed on the railcar with impact limiters for transport.

The transfer cask is lifted and positioned on the NAC-STC. The canister is lowered into the cask. The transfer cask is removed, and the NAC-STC lid is replaced and bolted.

Figure 9. Transport Preparation and Loading

mitment that would be necessary to develop a new fuel management concept especially for naval fuel. The only changes that need to be made to the cask design developed for commercial fuel management are optimization of the physical dimensions of the containers and overpacks and verification that the characteristics of the fuel are bounded by the nuclear analyses supporting the cask design. Neither of these are expected to present a problem.

The advantages of using multi-purpose casks for naval fuel are similar to those that apply to commercial fuel: flexibility and economy. Storage capacity can be provided as needed without the necessity of constructing an engineered facility such as a water-filled pool. When it is required, the spent fuel can then be transported for reprocessing or to another storage site, without the need for further fuel handling operations. Once a canister is unloaded, or sent away for disposal, the storage and transport overpacks previously used for that container can be reused.

Following the decision of the current United States administration to discontinue the reprocessing of submarine fuel, NAC is in discussion with the United States Navy concerning the use of its UMS for the management of United States naval fuel.

III. LOW- AND MEDIUM-LEVEL LIQUID AND SOLID WASTES

CHAPTER 7
APPLICATION OF IVO INTERNATIONAL LTD.'S "MOBILE NUCLIDE REMOVAL SYSTEM" (NURES) FOR LIQUID RADIOACTIVE WASTE TREATMENT AT THE PALDISKI NAVAL TRAINING CENTER, ESTONIA AND AT THE REPAIR AND TECHNOLOGY ENTERPRISE RTP "ATOMFLOT," MURMANSK, RUSSIA

JUHANI JOHANSSON, AAGE LAHTINEN, ESKO KINNUNEN, JUHAN TYRVÄINEN
IVO International Ltd.
Vantaa, Finland

1. Abstract

IVO International Ltd. has developed a mobile NUclide REmoval System (NURES) that allows for very effective purification of liquid radioactive wastes. The first section of this paper describes major characteristics of the facility.

In 1995, the NURES facility was taken to Estonia to purify the liquid waste that had accumulated from operating nuclear reactors at the Paldiski naval training center. About 760 cubic meters of liquid waste were purified, mainly from cesium. Only one 12-liter capacity filter was needed to treat the liquid. An estimated activity of 11.1 GBq (0.3 Ci) was collected in the filter, which was removed from the NURES facility, buried in a concrete container, and disposed of for long-term storage. This paper summarizes that project.

Another mission to treat liquid radioactive waste with the NURES facility involves transporting it to the RTP "Atomflot" enterprise in Murmansk. The project is called "ARKTIKA-M," and it includes purification of about 300 cubic meters of nuclear reactor cooling waters from the Russian icebreaker fleet operating in the Arctic seas. To this end, IVO International Ltd. and the RTP "Atomflot" signed a cooperative contract in May 1996. An outline of the project plan is presented in this paper.

2. The "NURES" System

2.1 MAIN CHARACTERISTICS

There is no other comparable method in the world at present which could compete with NURES in separating cesium from water. The separation principle is based on

71

the use of a selective inorganic ion exchange material, which very effectively separates cesium nuclides from liquid wastes. NURES also includes sophisticated filtering systems which remove other radionuclides such as cobalt, strontium, and cerium.

NURES is installed in a steel container six by 2.5 by 2.7 meters, which makes it compact and easy to transport (See Figure 1). Its transport weight is approximately seven tons.

The first step in practical waste treatment work is to transport the NURES facility to the waste site. In most cases, this can be easily achieved by truck or ship. The facility can be installed, preferably, near a waste tank in which the liquid to be treated is collected.

Because of a high degree of automation, operating NURES is relatively easy. It includes pumping the liquid from the waste tank through prefilters, the ion exchange columns, and other filters to a collecting tank (See Figure 2). In the course of the pumping process, the mechanical particles and the nuclides are captured in the filters. After a certain treatment period, the filters are replaced with new ones. The treated water in the collecting tank must meet international standards before it can be released. The used filters are deposited in concrete containers for disposal in solid waste storage sites.

Continuous control of passing nuclides is provided by automatic monitoring. The typical capacity of NURES is five to ten cubic meters per day of purified water.

3. Purification of Liquid Radioactive Wastes in Paldiski

3.1. SUMMARY

This report summarizes the work done by IVO International Ltd. (IVO IN) in decommissioning the Paldiski former submarine training center in Estonia.

IVO IN's role was to act as a contractor to the Paldiski Site Commander, who had been empowered through an Estonian-Russian agreement on decommissioning the Paldiski training center, and to execute the tasks specified in that agreement. IVO IN's performance during the contract was greatly helped by the cooperative attitude of the Russian site organization.

To begin the work, IVO IN set up a project to carry out the agreed waste treatment work, which finally resulted in successful purification of 760 cubic meters of liquid radioactive waste.

The technological basis for the work was a specific Nuclide Removal System (NURES) developed by IVO IN. Cesium (^{137}Cs) was the only nuclide to be removed from the liq-

Figure 1. Steel Container for NURES System

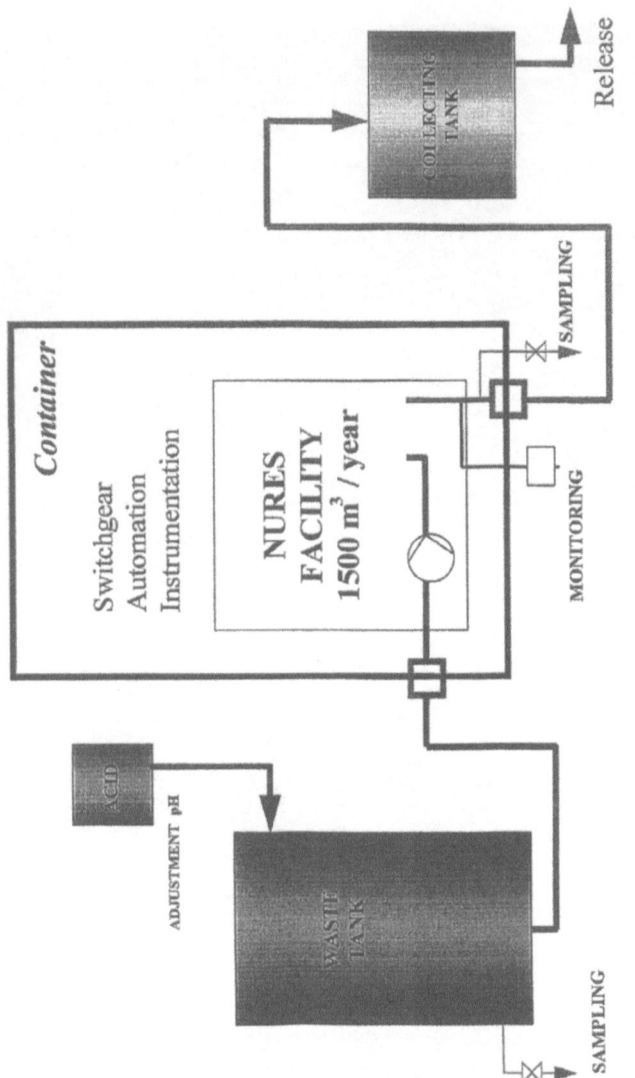

Figure 2. Scheme of Operations

uid waste. This was based on a temporary standard for water quality issued by the Estonian government.

An excellent decontamination factor of approximately 3000 was achieved in the waste treatment. The total collected activity was about 11.1 GBq (0.3 Ci), originating mainly from the cesium radionuclide, which was separated in a small filter with a volume of only 12 liters. The used filter was disposed of by sealing it in a concrete container designed for long-term storage. A remarkable result was that the reduction factor, defined as a volume ratio between the liquid and the concrete container, exceeded 400.

3.2. BACKGROUND OF THE PURIFICATION PROJECT

3.2.1. Contract Preparations

IVO IN's participation in the decommissioning of Paldiski was informally discussed in the summer of 1994 by the Estonian authorities and the Site Commander. It was then welcomed by both sides. Before signing a contract, the following questions had to be resolved:

- feasibility of IVO IN's NURES technology for treatment of the waste in Paldiski;
- feasibility of a Finnish company, IVO IN, as a contractor; and
- availability of funds to finance the project work.

The feasibility of IVO IN's technology for treatment of the waste was demonstrated by the results reached at the Loviisa nuclear-powered submarine (NPS), Finland, where cesium removal from evaporator concentrates using the NURES principle is a well-established practice.

Further confirmation of the NURES feasibility was received from the results of the Paldiski liquid waste analyses which indicated that the NURES treatment is very well suited to the purification task.

The temporary cleanliness standards issued by the Estonian authorities prior to the signing of the work contract indicated clearly that the release limits for treated liquid could be met by using the NURES facility.

In October 1994, a Paldiski International Expert Review Group (PIERG) meeting was held in Helsinki. During this meeting, IVO IN presented the NURES purification technology to representatives of the Russian Navy. As a result, IVO IN was welcomed to render purification services in Paldiski.

In the course of the above-mentioned PIERG meeting, the Ministry for Foreign Affairs of Finland decided to provide a major part of the financing for the purification.

The Helsinki PIERG meeting also included this work as Project C4 in the list of projects to be implemented under the auspices of PIERG.

3.2.2. Purification Contract

Paldiski Object's Site Commander, Rear Admiral A.V. Olhovikov, and IVO IN signed the contract on 21 December 1994. The Estonian authorities defined temporary release limits for purified waters, which were annexed to the contract.

The scope of IVO IN's services included:

- purifying the liquid radioactive waste in tank numbers one, five, and six until the remaining volume of liquid waste is roughly 30 cubic meters or until the bottom sludge level is reached (See Tables 1 and 1.1: Inventory of the storage tanks);

- purifying the liquid radioactive waste in tank number 15_2 to create space for the intermediate storage of the liquid waste entering the purification treatment; and

- delivering a concrete storage container for the used cesium filter columns and embedding the filter columns in it. At the time the contract was signed, the assumed, but not confirmed, volume to be purified was approximately 540 cubic meters.

3.2.3. Financing the Project

IVO IN and the Ministry for Foreign Affairs of Finland signed a contract on financing the purification project.

3.2.4. Other Parties Involved

The Estonian authorities, although contractually not directly involved in the purification work, exercised control over the work by setting temporary standards and by analyzing IVO IN's purification results. The Ministry of Environment of Estonia issued permits to Paldiski Object to release the purified water to the Gulf of Finland. Estonian laboratory services provided valuable assistance in characterizing the tank contents and monitoring the performance of the NURES facility.

Temporary standards, valid for the time of IVO IN's purification work in Paldiski, were announced on 16 December 1994 in Decree No. 465 of the Estonian government. Allowable limits were set at a level which was higher by a factor of 50 than the permissible atmospheric concentration (DK_B) in the Russian Radiation Safety Norms (NRB-)76/87. The corresponding values were: tritium 7.4×10^6 becquerels per liter (Bq/l), cobalt-60 1000 Bq/l, strontium-90 740 Bq/l, and cesium-137 1000 Bq/l.

IVO IN's work was regularly reported and discussed in consecutive PIERG meetings.

3.3. RESULTS OF THE PURIFICATION WORK

3.3.1. Characterization of the Tank Contents

The contract stipulated that reliable information on the characteristics of all tank contents be obtained and the purification scheme adjusted accordingly in order to avoid complicating the future solidification of tank bottom sludges.

TABLE 1. Paldiski Wastewater Purification: Inventory of Storage Tanks

Tank Number		Total Volume (m³)	Volume of Waste (m³)	Radionuclides in Water and Sludge (Bq/Kg)						Sediment Type
				137-Cs	134-Cs	60-Co	54-Mn	90-Sr	3-H	
1	W	400	111	*)11,800 **)14,500	38	3.71		230 200	145,100 145,800	
	S		ca. 3	***) 400,000		32,000		1,960,000		coagulant
2	W	400	19	50,300	120			40	165,000	
	S		ca. 1	380,000		20,000		4,870,000		coagulant
3	W	400	29	31,200				60	105,000	
	S		ca. 1	493,000		10,000		3,045,000		coagulant
4	W	400	24.5	23,000	60			80	82,000	
	S		ca. 1	427,000		13,000		1,500,000		coagulant + organic pitch or shellac
5	W	400	115 [1]	359 9100	0.83 60	2.0 20		230 2460	247,000 160,000	
	S ****		ca. 15–50 ?	3,980,000		740,000	137,000	610,000		ion exchange resin
6	W	400	125 [1]	239 3900	0.67	1.6 50		960 220	130,000 232,000	
	S ****		ca. 30 ?	208,000		1,280,000		61,000		sand and ion exchange resin

W=water phase   S=sludge phase

Remark: The samples of tanks No. 5 and 6 may have been erroneously mixed with each other in the 1994 sampling.

Analyses made: *) before 1995 for water   **) January 1995 for water   ***) January 1995 for sludge
****) Samples taken after mixing of water phase above and sand filters
[1] Depth measurement to the top surface of ion exchange resin layer

78

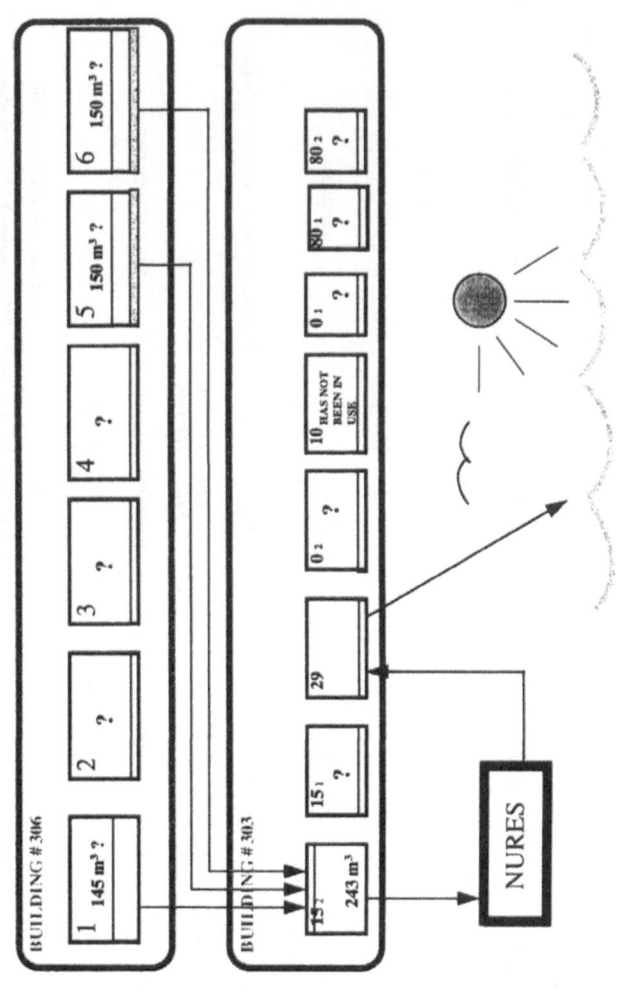

However, at the time the contract was signed, knowledge of the content of the tanks was rather general, as it was mainly based on the earlier information of Russian organizations, with some supplemental water analyses taken for IVO IN.

The problem for proper characterization was to provide access to the tanks which were considered to be hermetically sealed except for piping connections. A drilling effort through the roof slabs had already been planned toward the end of January. It then appeared that there were manlocks on the tanks under the floor finish of the waste building. The manlocks of tank numbers one through six were freed from this finish and opened. After that, both water and sludge phases were sampled in all tanks in the storage building. The results of these analyses and volume measurements are shown Table 1.

3.3.2. Purification Work
IVO IN's purification work officially began on 11 January 1995. The work, consisting of four separate purification rates, stretched over a time span of seven months, ending on 23 August 1995. A total of 760 cubic meters of waste liquid was treated. The work proceeded principally according to the sequence of schemes presented in Table 2.

TABLE 2. Paldiski: Daily Purification Rates

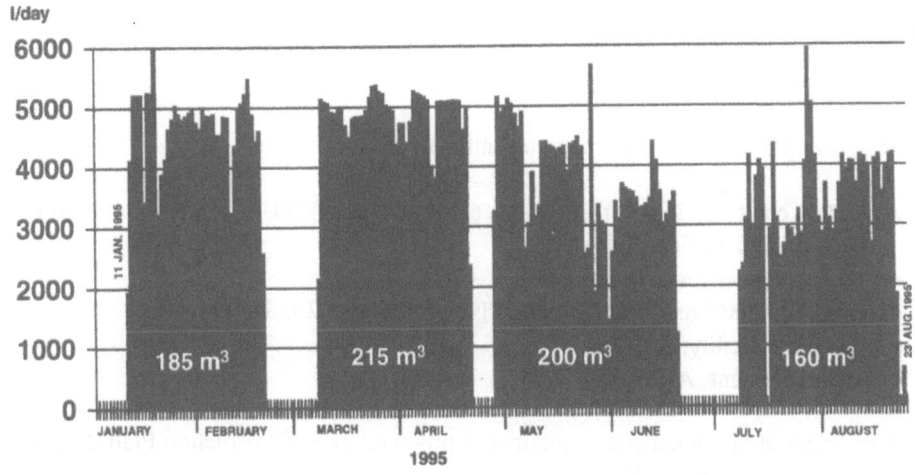

The first tank to be worked with was number 15_2, partially because it was needed as an intermediate storage tank for further work, but also in order to gain time to collect more information on tank numbers one, five, and six.

Purified water was given over in four separate installments. The official results of the water analyses at the takeovers are shown in Table 3.

According to the results of the first three rates, a decontamination factor of approximately 3000 was achieved. The fourth rate, where an essential increase in the radioac-

80

tivity level can be observed, may not be considered as representative for the reasons discussed in section 3.4.

TABLE 3. Official Results of Water Analyses Takeovers

	1. Release 27 February 1995	2. Release 25 April 1995	3. Release 4 July 1995	4. Release 9 September 1995	Total Release
Volume (m³)	185	215	200	160	760
^{137}Cs	0.58 Bq/kg	< 0.3 Bq/kg	< 0.3 Bq/kg	93 Bq/kg	0.015 GBq (0.0004 Ci)
^{90}Sr	50 Bq/kg	550 Bq/kg	647 Bq/kg	242 Bq/kg	0.296 GBq (0.008 Ci)
^{60}Co	0.65 Bq/kg	1.1 Bq/kg	1.8 Bq/kg	5.2 Bq/kg	0.002 GBq (0.00004 Ci)
^{3}H	76.5 kBq/kg	133.8 kBq/kg	123 kBq/kg	48.8 kBq/kg	75.36 GBq (2.06 Ci)

3.3.3. Deposition of the Cesium Filter and Other Contaminated Materials

Concrete Container. For depositing the used cesium separation filter, a special concrete container, shown in Figure 3, was provided. The container was manufactured by a Finnish company, Myllylän Betoni Oy, according to IVO IN's engineering design. The container was transported to Paldiski on 24 August. The spent cesium filter was placed in the container, which was sealed with an in-situ-concreted lock. This work was performed by IVO IN's specialists.

The main parameters of the concrete container are given below:

- safety class: EYT (according to the classification of STUK, Finnish Center for Radiation and Nuclear Safety)
- operating/design temperature: 0.... -40/+20°C
- radiation level on the container surface: maximum 2 mSv per hour
- concrete quality: K60-1
- reinforcement: AISI 304

A description of the concrete container is presented in a Construction Plan document, YDIN-GM9-1 (in Finnish).

An identification code, 01-NURES 21.9.1995, was painted in black on one side of the container.

The used mechanical filters, altogether 48 pieces with an approximate size of six centimeters in diameter and 60 centimeters in length, are stored in a 200-liter steel barrel, coded with 10-NURES 21.9.1995. Other leftover materials, e.g. package covers and clothing, are stored in another 200-liter steel barrel, coded with 11-NURES 21.9.1995.

All three waste containers are stored in the building in which the NURES facility was in operation. This is in complete accordance with the purification contract.

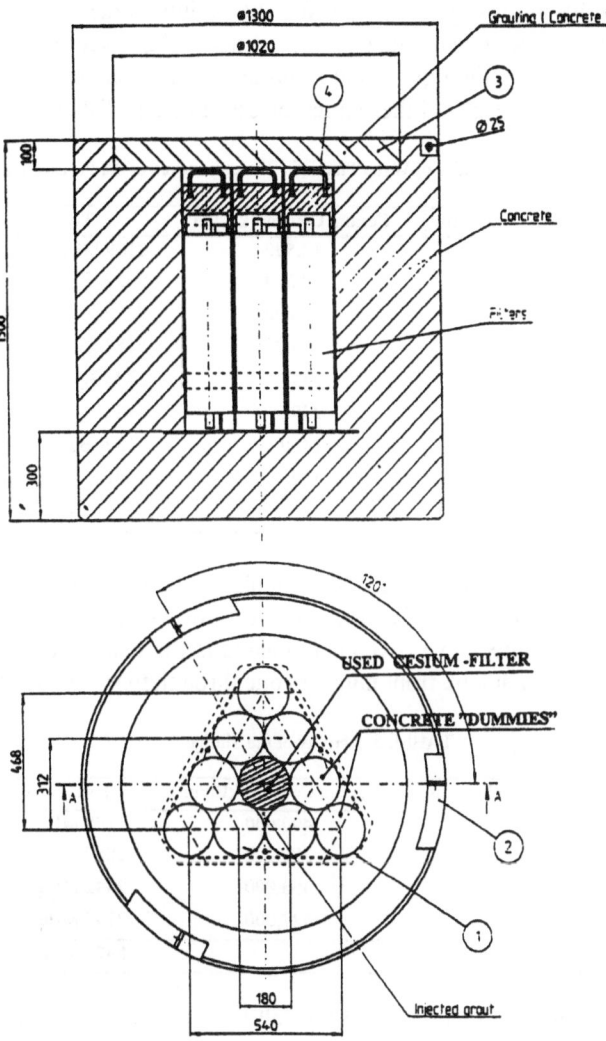

Figure 3. Paldiski: Concrete Container for Final Disposal of Used Cesium Filter

Level of Radioactivity in the Used Cesium Filter. No activity measurements of the deposited cesium filter were considered necessary. The following provides an approximate estimated range of the total activity. The bases for the estimation are the results of the radioactivity level measurements conducted at the input and output of the NURES facility during the purification work. The remaining activity in the sludge and in the sand filters constitutes an uncertainty factor to this estimation. This is especially true because the samples taken under the sand filters were found to have less cesium activity by a factor of 30 than the samples taken above the sand filters.

It is assumed in the following estimation that the cesium values from tank numbers five and six are the values taken under the sand filter. This corresponds to the actual pumping-out procedure of the liquid. No trapping of other nuclides in the filter was taken into account. The results of the estimation are shown in Table 4.

TABLE 4. Total ^{137}Cs Activities of Water Phases in Tanks Involved in the Purification Process

Tank Number	Purified Volume (m³)	Cesium Content (kBq/m³)	Total Cs Activity	
1	145	14,500	2109 MBq	(0.057 Ci)
2	19	50,300	962 MBq	(0.026 Ci)
3	29	31,200	925 MBq	(0.025 Ci)
4	25	23,000	555 MBq	(0.015 Ci)
5	130	359	37 MBq	(0.001 Ci)
6	130	239	37 MBq	(0.001 Ci)
15_2	240	1460	333 MBq	(0.009 Ci)
80_1	40	150	0 MBq	(0 Ci)
	758		4958 MBq	(0.134 Ci)

Additional activity coming from the sludge of tank numbers one through four, assuming that the cesium in these tanks was in soluble form when the tanks were washed and emptied to tank number five (five cubic meters is the maximum estimated value for sludge volumes in each tank from numbers one through four), is shown in Table 5.

TABLE 5. Activity of the Sludge

Tank Number	Purified Volume (m³)	Cesium Content (kBq/m³)	Total Cs Activity (Ci)	
1	5	400,000	1998 MBq	(0.054 Ci)
2	5	380,000	1887 MBq	(0.051 Ci)
3	5	493,000	2442 MBq	(0.066 Ci)
4	5	427,000	2146 MBq	(0.058 Ci)
	20		8473 MBq	(0.229 Ci)

A summary of the these estimations gives a result that the maximum activity of the filter cartridge was 13,320 MBq (0.36 Ci).

Decontamination of the NURES Facility. The NURES facility had to be decontaminated for the purpose of transporting it from the Paldiski site. After finishing the purification work, the facility was connected to a closed washing circuit, bypassing the cesium filter. The main phases of the decontamination procedure are given below:

- Approximately 200 liters of four percent NaOH solution were pumped through the system.

- The system was flushed with approximately 800 liters of demineralized water.

- Final treatment included flushing with approximately 200 liters of a four percent citric acid solution.

After the first decontamination cycle, the activity levels in different locations of the facility were measured in order to check for a need to repeat the washing cycle. However, the result was found to be satisfactory, and the inspectors of the Russian radiation control authorities supplied a certificate on the successful decontamination.

Based on the cleanliness certificate of the facility, the Estonian health authorities granted permission for transportation of the NURES facility within the territory of Estonia. In addition, the Finnish Center for Radiation and Nuclear Safety (STUK) granted the permission needed to import the facility to Finland. The facility was transported from Tallinn to Helsinki on 21 September 1995.

3.4. EXPERIENCE FROM PRACTICAL WORK

No essential problems were encountered after the contract had been signed and the work begun.

The NURES purification facility is fully automated and, therefore, only a casual attendance by the personnel was needed. However, disturbances occasionally occurred. Normally during the day, these were soon remedied by the Russian site organization. However, at night and during special weekends, some switch-offs due to these disturbances proceeded unnoticed for a period of up to one half day or one day.

At the beginning, the facility was found to be rather sensitive to the exceptional instability of the external power supply. Most switch-offs were due to this factor. Some pressure alarm settings of the system required experimentation, as well, before smooth operation of the facility was achieved.

Delivery of waste water from tanks other than number five to tank number 15_2 caused pumping problems. Therefore, the waters from the storage building were pumped through tank number five. There were some concerns that the mud which was emptied from tank numbers one through four to tank number five would migrate through the sand filter and enter the purification process. This assumption turned out to be true because during the third purification period, a rapid increase of pressure difference over the mechanical pre-filters was observed.

When the filters were changed, they were found to be heavily contaminated with a black material. Furthermore, when the purification work was about to be finished, similar findings of pressure rise and the contamination of mechanical prefilters were observed. Additionally, an oil film was observed on the surface of water in the housing of the mechanical prefilters.

3.5. CONCLUSIONS

The work performed by IVO IN was carried out as a part of the total decommissioning of the Paldiski waste storage and waste treatment buildings.

IVO IN's role included purification of 760 cubic meters of liquid radioactive waste. The main radionuclide removed was cesium. It was collected in one ion exchange column with a volume of 12 liters. The spent column was deposited in a concrete container for long-term storage. The entire job was completed by using the NURES purification facility, which was transported to Paldiski.

The purification project proved that problems of purification of liquid radioactive waste can be resolved quickly and economically by using the mobile, high efficiency, selective nuclide removal NURES facility.

4. LRW Treatment Project at RTP "Atomflot," Murmansk, Russia

4.1. INTRODUCTION

In the course of the past few years, a considerable volume of liquid radioactive waste (LRW) has accumulated in storage tanks in the Murmansk region of Russia. This is due to the fact that the discharge of LRW into the Arctic seas was reported be stopped in 1992 and capacity to treat all the generated LRW was not available.

The major part of the LRW originates from operation and maintenance of the surface and underwater vessels powered by nuclear propulsion.

Despite local treatment efforts, the storage tanks are reported to be full of LRW with an eventual risk of spilling over or leaking out into the environment due to corrosion of the tanks.

The organization committed to LRW treatment in the Murmansk region is the RTP "Atomflot" (Repair and Technology Enterprise "Nuclear Fleet," Murmansk). This enterprise has successfully treated LRW using Russian technology. The Association for Advanced Technologies, ASPECT, Moscow is the company which provides the know-how of the Russian LRW technology. At this moment, the RTP's facilities are under major renovation and modernization. Consequently, there is practically no LRW treatment taking place in Murmansk now.

To tackle the problem of the accumulating LRW, a Finnish company, IVO International Ltd., proposed to the Russian side a test of the services of a mobile "NURES" radionuclide removal system which recently performed an LRW treatment mission at the former military base of Paldiski, Estonia with outstanding results. Conclusions about the feasibility of the NURES facility at the RTP "Atomflot" will be drawn based upon test results.

A contract between IVO International Ltd. and the RTP "Atomflot" was signed on 8 May 1996 in Murmansk. This contract includes delivery of the NURES facility to the RTP "Atomflot." After installation, a volume of 300 cubic meters of LRW originating from the nuclear icebreaker fleet will be treated. This is expected to take a few months. Afterwards, the facility will be adjusted to allow for treatment of LRW with higher salt content and more complex radionuclides. The project is financed by the Ministry for Foreign Affairs of Finland.

4.2. OBJECTIVES

The first objective is to deliver from Finland to Murmansk the transportable, high efficiency NURES LRW treatment facility, which allows for removal of radionuclides from liquid wastes and disposal of the removed nuclides in solid form, which, in turn, results in the significant reduction in volume of the remaining liquid waste.

The primary objective is to treat about 300 cubic meters of primary circuit cooling waters of the icebreaker reactors. The second project phase includes adjusting the facility to treat LRW with higher salt content and more complex radionuclides and continuing the treatment work.

Another important objective is to include Russian LRW treatment technology in the NURES facility. This would provide an increased treatment capacity and more independent operation of the facility by the RTP "Atomflot."

4.3. LRW TO BE TREATED

The LRW to be treated belongs to the first group (primary circuit reactor waters) and the second group (radioactive contaminated solutions) with a radioactivity in the range of 37 to 370 kBq (10^{-5} to 10^{-6} Ci) per liter. The important radionuclides to remove are cesium and strontium.

The target is to treat the LRW to meet international and Russian release limits. The Russian norms are *Radioactive Safety Standards NRB-76/87*, *Basic Health Regulations OSP-72/87*, and *Health Regulations for the Handling of Radioactive Wastes SPORO-85*.

IVO IN expects to reach decontamination factors (DF) on the order of 10,000 to 20,000.

The treatment capacity of NURES is up to ten cubic meters per day. In this project, the target volume to be treated is about 300 cubic meters. Depending on the first treatment results, this volume can be expanded considerably.

4.4. NURES FACILITY

The NURES facility is a modernized version of the Nuclide Removal System which was originally developed by IVO International Ltd. and which has been operated successfully at such places as the Loviisa NPS, Finland and the former military base of Paldiski, Estonia. Modernization is needed to adjust the system to meet the specifications of the LRW to be treated at the RTP "Atomflot."

The treatment process is based on filtration and ion exchange, which very effectively separate particles and radionuclides from water. The general layout of the facility is shown in Figure 4.

First, the liquid to be treated is collected in a large waste tank near the facility (See Figure 2). After pH adjustment, the liquid is pumped through prefilters and ion exchange columns to a collecting tank. In the course of the pumping process, the mechanical particles and nuclides are captured in the filters and columns, respectively. After a certain treatment period, the saturated filters and columns are removed from the system and replaced with new ones. The used filters and columns are deposited in concrete containers for disposal in solid waste storage facilities. The activity of the treated water is monitored and recorded continuously. Samples are taken periodically. If the activity with respect to each individual nuclide is below the allowed level according to international and Russian regulations, the water can be released to sea. If not, the treatment will be repeated.

The facility will be installed inside a building reserved for the treatment of radioactive waste at the RTP "Atomflot." The storage tanks for the waste liquid and the reception tanks for the treated liquid are available in this building.

4.5. TIME SCHEDULE

The total length of the project is eleven months starting from the signing date of the contract—8 May 1996. Modernization of the facility and delivery to Murmansk will occupy the first three months. Licensing, installation, test runs, training of the RTP's personnel, treatment of the LRW, and decommissioning of the facility will take about eight months. In the course of the work, new plans to continue treatment of the LRW will be drawn up.

4.6. PARTICIPATING ORGANIZATIONS AND DIVISION OF WORK

Below, the major tasks of the participating organizations are outlined:

IVO International Ltd., Finland
- Modernization and delivery to Murmansk of the NURES facility.
- Supervision of installation and commissioning of the facility.
- Supervision and participation in the treatment work of the LRW.

Figure 4. General Layout of the NURES Facility

Figure 5. Computer Image of the NURES Facility Installed on the Deck of the Ship *Serebryanka*

- Training of the RTP's specialists and technical support.
- Project management.

RTP "Atomflot," Murmansk
- Purification site, services, and materials.
- Installation and commissioning of the facility.
- Operation of the facility.
- Disposal of the spent filters and columns.
- Permissions and licenses.

4.7. CONCLUSIONS

The Russian and Finnish enterprises have undertaken to jointly treat 300 cubic meters of liquid radioactive waste (LRW) originating from the cooling circuits of the ice-breaker fleet's reactors in Murmansk.

The joint project between IVO International Ltd. and the RTP "Atomflot" was started by signing a contract on 8 May 1996. The project includes an LRW treatment period of six months, as a result of which the purity of the treated water shall meet international and Russian norms.

The project will be financed by the Ministry for Foreign Affairs of Finland.

Figure 5 shows a computer image of the NURES facility installed on the deck of the ship *Serebryanka*.

CHAPTER 8
THE DEVELOPMENT OF A MODULAR PLANT FOR PROCESSING LIQUID RADIOACTIVE WASTES FROM THE NUCLEAR FLEET*

"Korvet" Project

R.A. PENZIN
"Aspect" Association
V.S. SHEPTUNOV
NPP "Biotechprogress"
B.M. LESOHIN
NPO "Zvyozdochka"
B.K. BULYGIN
NPP "Ecoatom"
A.A. SHVEDOV
VNIPIET

1. Justification for the Project

In order to reprocess liquid radioactive wastes of various chemical compositions, an integrated technology ensuring that the sanitary and radiation standards of the Russian Federation for purified solutions are met will be required.

For this, the stipulated limiting condition for all processes is that treated radioactive wastes must be in the form of cement blocks which satisfy all the parameters required by Sanitary Regulation for Handling Radioactive Wastes (SPORO)-85 and purified water with a total content of radionuclides of less than ten curies per liter, salt content no higher than 200 to 300 milligrams per liter, and other impurities in accordance with the requirements for fishery waters.

The current redesigned installation for processing liquid radioactive wastes (LRW) of the nuclear fleet at Technological Repair Facility (RTP) "Atomflot" with a projected production rate of 5000 cubic meters per year will comprise a complete reprocessing cycle, including the assembly for hardening brines and waste sorbents in reinforced concrete protective containers.

The installation is intended to reprocess all three types of liquid wastes:

---

* Translated from the Russian by Sanoma Lee Kellogg, AAAS

E.J. Kirk (ed.), Decommissioned Submarines in the Russian Northwest, 91–94.
© 1997 *Kluwer Academic Publishers.*

- type 1—low-salt solutions (circuit waters);
- type 2—decontaminated solutions (with salt content less than or equal to six grams per liter); and
- type 3—saline wastes (specific wastes from the Navy which are generally a mixture of solutions and seawater with a salt content between seven and 20 grams per liter).

These wastes include nonstandard wastes formed in the operation of nuclear submarines with the technical services of the Navy.

All projected indices of the redesigned plant are based on initial total activity of LRW in the interval of 10^{-5} to 10^{-7} curies per liter. However, in reality, wastes produced by the Navy can have activity levels on the order of one or two times higher. Thus, the inspection of tanks with LRW on the TNT-12 tanker indicated an excess of activity by more than one order above the projected values in six out of eight tanks. In a series of instances on naval vessels, so-called "nonstandard" liquid radioactive wastes are formed which necessitate the use of a special, relatively complicated and labor-intensive reprocessing technology that significantly raises the cost of the processing and leads to the formation of a large quantity of solid wastes. In order to avoid all of these negative consequences and simplify the reprocessing of LRW on the regional base plant in the city of Murmansk, the implementation of the "Korvet" project is proposed in addition to operation of the base plant.

2. Project Description

The essence of the given project lies in the creation of compact, mobile installations for purifying liquid radioactive wastes. These installations are secure and simple to operate. They perform well in the form of a module with the dimensions of a standard marine container. They are equipped with all of the systems that maintain their operation in both the coastal version and the floating version when installed on a special floating facility, and they must be connected to a source of electricity (380 volts). Their primary purpose is to reprocess LRW at the site of its formation. This will allow for the elimination of the production of LRW of the third type (as defined above), thereby significantly simplifying and increasing the security of the base LRW reprocessing system in the city of Murmansk.

The developed modules are based on current sorption-membranous technologies for the purification and concentration of LRW, which have performed well in practice in pilot LRW reprocessing operations with civilian and military nuclear fleets in the city of Murmansk at RTP "Atomflot" and at five objects of the Pacific Fleet.

The schematic diagram of LRW purification in the "Korvet" module will consist of passing the solutions first through special selective inorganic sorbents and then through a reverse osmosis module. To finish, the solutions are additionally purified with the help of special zeolites and ion exchange resins. Radioactive wastes produced in the process of the operation (spent sorbents and brines) with a concentration of 60 to 80 grams per liter are subject to final reprocessing and reclamation at installations for concentration and subsequent hardening.

In addition, the plant provides a sorption-microfiltration system for processing mild brine solutions. This system ensures the greatest level of radionuclide concentration in the solid phase.

The hardened product is ready for final burial in the type of standard reinforced concrete containers used in the Murmansk project.

3. Basic Technical Characteristics

3.1. DIMENSIONS

The overall dimensions of the constructed module are not greater than width 2.7 meters by length 12.0 meters by height 3.2 meters.

3.2. TYPE OF PROTECTION

The principal protective elements are lead and concrete.

3.3. PRODUCTIVITY

The plant has a production rate of $0.3 \div 1.0$ cubic meters per hour.

3.4. TYPE OF SOLUTIONS REPROCESSED

The plant processes all types of solutions, including non-saline, decontaminated, and saline with a salt content of up to 35 grams per liter.

3.5. PURIFICATION INDICES

The plant's purification indices range from 10^{-4} curies per liter to 10^{-10} curies per liter (less than 10^6).

The content of individual radionuclides is less than required by Radiation Safety Norms (NRB)-76/87.

The salt content of waste solutions is less than 0.1 grams per liter.

3.6. EXTENT OF RADIONUCLIDE CONCENTRATION

The extent of the radionuclide concentration is:

- 5×10^{-2} curies per liter in solid phase; and
- 10^{-4} curies per liter in brine.

3.7. CONTROL SYSTEM

The installation is operated remotely with duplication by manual control.

4. Conclusion

The launching of a pilot prototype equipped with all of the expendable materials necessary for carrying out tests is planned for 1997.

The creation of compact modules for processing LRW is part of the regional technical policy for handling radioactive wastes. The implementation of this project will significantly increase the security of the entire LRW reprocessing system and completely eliminate the unauthorized discharge of LRW into the sea, as the more dangerous LRWs will be decontaminated at their production site. Furthermore, through an insignificant increase in general expenditures for the creation of the complex, the implentation of this project will significantly lower operating costs.

One of the major benefits of the proposed idea for a regional system of LRW reprocessing is the significant reduction in solid radioactive waste (in cement blocks) awaiting interment. Simple calculations for the output of the installation at RTP "Atomflot" indicate that the quantity of solid radioactive wastes there will be reduced by no less than two times. In view of the final burial of solid radioactive wastes at Novaya Zemlya planned in the regional program, this indicator alone will recoup all expenditures for building the "Korvet" project's installation.

The significant experience of Russian firms—"Aspect" Association, NIPTB "Onega," NPP "Biotechprogress," NPP "Ecoatom," and RTP "Atomflot"—in reprocessing LRW of various types will be utilized in the course of implementing this project.

The implementation of this project is in complete accordance with the statements in Article 18 of the "State Technical Policy on Handling Radioactive Wastes."

CHAPTER 9
LOW AND MEDIUM RADIOACTIVE SOLID WASTE

Concept For the Development of a Waste Treatment Facility

SERGE MERLIN
SGN Reseau Eurisys
St. Quentin Yvelines, Cedex, France

1. Background

Environmental pollution risks in northwestern Russia stem from both military and civilian nuclear activities. The radioactive sources present in this region are of three main types :

- Spent fuel from the operations of nuclear submarines and icebreakers currently stored either aboard floating bases or in laid-up submarine reactors or terrestrial storage sites.

- Radioactive sources dumped in the Kara and Barents seas, including a few submarine reactors sunk with their fuel, and waste from the operations of the above nuclear ships, which was dumped in the sea until 1993.

- Liquid effluents and solid wastes generated since 1993 and stored on various sites for subsequent treatment and conditioning.

The risks of polluting the environment depend on many factors, including total source activity, source dispersibility, and the number and quality of containment barriers around the radioactive material.

Liquid effluent treatment is currently handled within a three-party Russian/Norwegian/United States agreement. This project has been given high priority in order to have a treatment facility available before the current storage tanks are filled.

Solid waste treatment must also be addressed now for the following reasons:

- Some solid waste, particularly bulky waste, is stored outdoors and therefore entails the risk of radioactive release.

E.J. Kirk (ed.), Decommissioned Submarines in the Russian Northwest, 95–112.
© 1997 *Kluwer Academic Publishers.*

- Its contribution to the total source term is presently almost negligible but it should become more significant after spent fuel disposal.

- The quantity of waste generated is already significant and will continue to increase in the years to come as a result of both the operations of nuclear ships and, above all, the dismantling of nuclear submarines and ships. About 5,000 cubic meters of waste are expected to be produced per year, and it will require huge storage areas if it is left untreated.

The purpose of this paper is to propose a concept for a treatment facility for solid waste produced in the Murmansk region based on similar European facilities.

2. Existing European Facilities

Two centralized waste treatment facilities are described below. They are intended for solid waste and, to a lesser degree, for low- and medium-level liquids, generated by:

- so-called "small producers" (research centers, laboratories, hospitals, industry); and
- pressurized water reactor (PWR) and boiling water reactor (BWR) power plants.

The wastes concerned arise from normal operations and from dismantling operations. They include:

- burnable waste;
- metal waste, compactible or not;
- resins; and
- miscellaneous sludges.

"Small-producer" wastes could be compared to the wastes generated in Russia by the Radon Institutes, particularly the Murmansk Radon Institute in northwestern Russia.

Nuclear Power Plant (NPP) wastes could be compared to those from the operations of nuclear submarines and icebreakers.

2.1. THE CILVA FACILITY IN BELGIUM

The CILVA conditioning facility in operation on the DESSEL site in Belgium has the following main functions :

2.1.1. Process Functions (See Functional Diagram in Figure 1)
- Waste reception and lag storage.

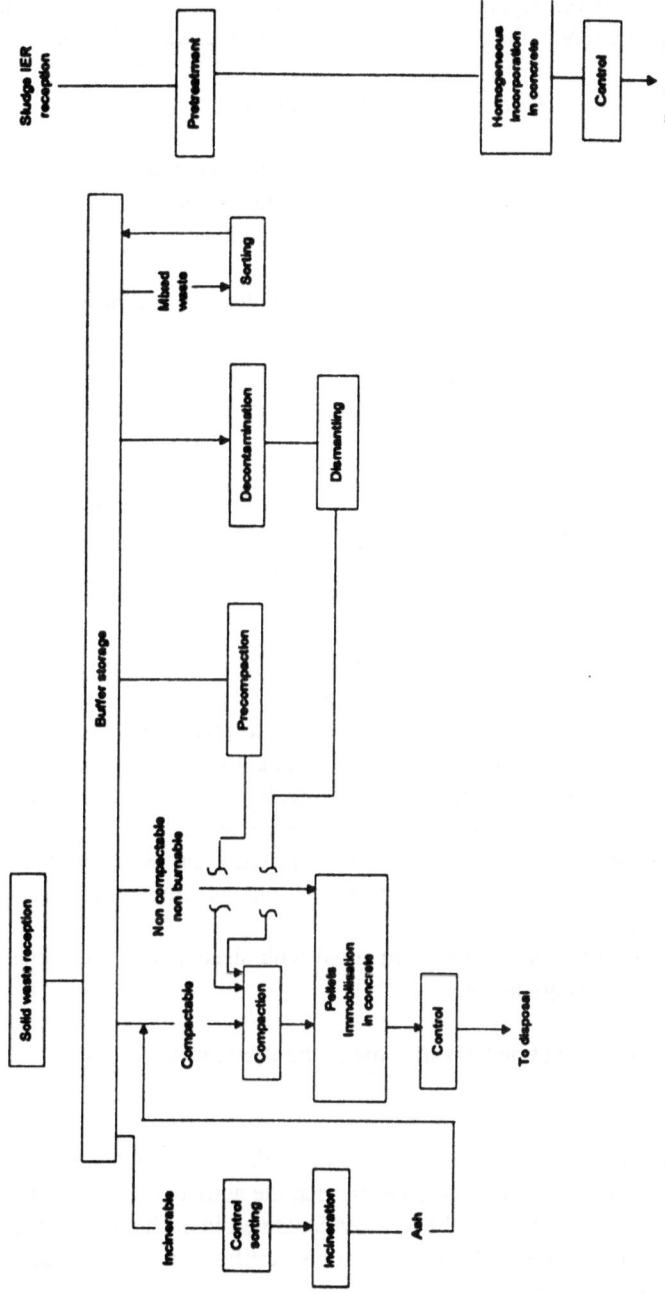

Figure 1. Overall Functional Diagram: The CILVA Facility

- Checking and, if necessary, sorting of burnable waste. Sorting is done at the production site. At CILVA, the only systematic operation is waste checking.

- Sorting of mixed waste (received in 200-liter drums).

- Precompaction of bulky wastes (greater than 200 liters) to reduce their volume and directly insert them into standard 200-liter drums (ventilation filters, activat-ed charcoal filters, etc.) for subsequent high pressure compaction.

- Incineration of combustible waste. Ashes are collected in 200-liter drums and super-compacted.

- Solid waste compaction, using a 1500-ton press. The resulting compacts are placed in 400-liter drums.

- Immobilization of the compacts and of non-compactible, non-burnable waste in a cement matrix (heterogeneous waste form).

- Encapsulation in a cement matrix of ion exchange resins, contaminated earth, and miscellaneous sludge (homogeneous waste form).

- Decontamination of cemented waste drums or bulky dismantling waste. This operation is performed with steam in a booth able to receive packages with maximum dimensions of length, three meters; width, two meters; and height, two meters. Large dismantling wastes are systematically decontaminated to remove the non-fixed surface activity prior to cutting them into pieces, whereas only off-standard cemented waste drums are decontaminated upon detection of non-fixed activity or activity trapped in grease during non-contamination checking.

- Dismantling of bulky waste in a room fitted with shears, saws and other cutting tools and ventilation equipment.

Figures 2,3,4,5, and 6 present pictures of some of the equipment.

2.1.2. Waste Identification/Tracking Function
Packages entering the facility or transferred through the facility are permanently tracked by means of bar code readers to ensure that the activity can always be located and to make a precise inventory of the activity contained in each package leaving the facility. Waste identification/tracking also permits certification of outgoing packages and acceptance of packages in final repositories.

2.1.3. Auxiliary Functions
Logistic support functions include:

- civil works;

Figure 2. Waste Preparation: Routing and Sorting Device for Incinerable Waste

Figure 3. Waste Preparation: 140-Ton Capacity Precompactor

Figure 4. Waste Preparation: Steam Jet Decontamination Device

Figure 5. Conditioning (Embedding of Heterogeneous Waste): Mixer for Non-Radioactive Grout Preparation

Figure 6. Conditioning (Embedding of Homogeneous Waste): Specific Mixer for Radioactive Waste Embedding

- building ventilation; and
- electricity and reagents distribution.

The facility is operated in day shifts, except for the incineration units, which are run round the clock in campaigns.

The capacity of the main units is:

- incineration: 100 kilograms per hour;
- compaction: 20 200-liter drums per day; and
- immobilization of compacts into cement: 14 drums per day.

2.2. THE ZWILAG CONDITIONING FACILITY

The centralized conditioning facility under construction on the Würenlingen site in Switzerland has the following main functions.

2.2.1. Process Functions (See Functional Diagram in Figure 7)

- Reception of solid and liquid waste.

- Checking of drum activity.

- Buffer storage.

- Sorting in a tight stainless steel glove-box of mixed waste received in 200-liter drums into:

    - burnable;
    - compactible; and
    - non-burnable and non-compactible waste.

- Dismantling of bulky wastes to reduce their volume and either insert them into 200-liter drums prior to compaction or decontaminate them. The maximum waste dimensions are four meters by two meters by two meters. These operations are carried out in a shielded box to minimize irradiation doses to workers. The box is fitted with appropriate dismantling tools (remotely operated plasma arc, cutting discs, etc.).

- Decontamination of metallic pieces to levels compatible with release for public re-use is performed using different methods such as high pressure water, vapor blasting, immersion in chemical baths with ultrasonic devices, and electrolytic processes. Decontamination equipment is installed in the dismantling box.

- Compaction using a high-force compactor (1500 tons) to reduce the waste volume by a factor of three. Compacts are collected in drums.

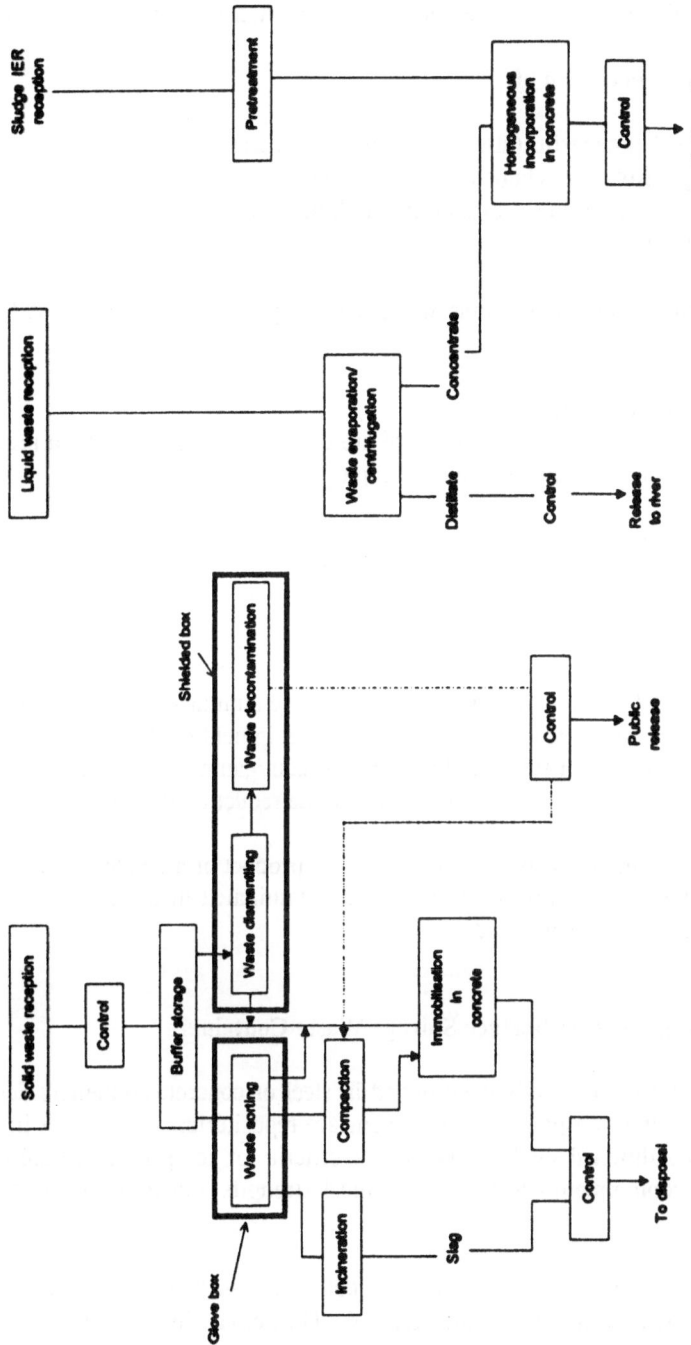

Figure 7. Overall Functional Diagram: The ZWILAG Facility

- Immobilization of compacts and non-compactible/non-burnable waste in cement.

- Encapsulation in cement of:

  - ion exchange resins (after pretreatment if necessary);
  - sludges from miscellaneous sources; and
  - concentrates from the evaporation of liquid effluents and decontamination solutions.

- Incineration of solid combustible waste using a plasma torch system with direct slag production.

2.2.2. Waste Identification/Tracking
The characteristics of the conditioned waste are obtained and certified by a procedure similar to that described in the previous example.

2.2.3. Auxiliary Functions
The same as for CILVA.

2.3. SUMMARY

The CILVA and ZWILAG conditioning facilities have similar functions, even though ZWILAG has more sophisticated decontamination equipment to allow material release for public reuse. They can treat the full range of beta/gamma wastes that are conditioned in high quality matrices with optimum volume reduction factors.

Other reference facilities involving a single type of product or a single technology can be quoted, for instance melting facilities for decontaminated metallic waste to be recycled as drums or biological shields.

3. Recent Developments in Surface Storage Waste Containers

Radioactive wastes are currently conditioned in steel or concrete containers designed for safe handling and transportation to final surface repositories. The matrix in which the wastes are conditioned shall satisfy specific criteria for acceptance in final repositories (such as containment capacity, mechanical strength, and resistance to thermal cycling).

A new type of container, the "high integrity container" or "durable confining envelope," has recently been developed and certified. The activity limits set for acceptance of the wastes in surface repositories remain unchanged but lower performance is required from the waste encapsulation/immobilization matrix since confinement of the radioactivity throughout the storage period is ensured by the envelope. Less efficient encapsulation matrices allow higher waste incorporation ratios.

Fiber-concrete containers belong to this new category. Two versions are available depending on the type of waste :

- cylindrical containers (Figure 8 depicts the CBF-C1 type container); and
- cubical containers (Figure 9 depicts the CBF-K type container).

Figure 8. Cylindrical Container CBF-C1 Type (Height: 1 Meter)

Figure 9. Cubical Container CBF-K Type

4. Definition of a Concept for the Treatment of Waste Generated in the Murmansk Region

This section does not intend to describe the facility to be built in the Murmansk region but rather summarizes the basic options to develop in order to devise a concept and proposes arguments for their selection. Figure 10 depicts a possible functional diagram for a Murmansk facility.

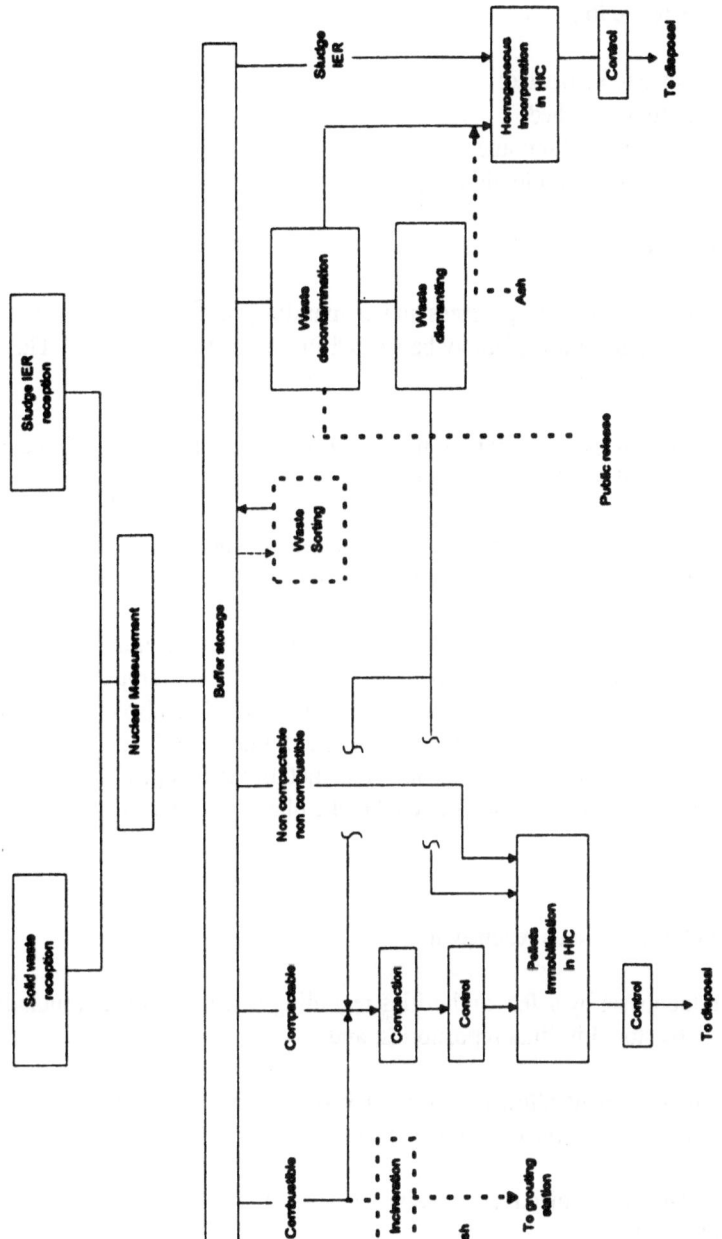

Figure 10. Possible Functional Diagram: Facility for Low- and Medium-Level Solid Waste Produced in the Murmansk Region

The basic options relate to :

- the type of waste to be treated;
- the objectives to be reached;
- the degree of centralization; and
- the utilization of fixed or mobile equipment.

4.1. TYPES OF WASTE TREATED

The question is whether operating wastes and dismantling wastes from nuclear submarines and icebreakers should preferably be treated in one or two facilities. The main criteria considered might be:

- the total treatment capacity, which may require several items of equipment that are easier to install on several sites; and

- the future of dismantling wastes, which wastes are usually produced in large quantities and can be recycled in the nuclear industry or even the non-nuclear industry after very efficient decontamination to reduce their specific activity to a few becquerels per kilogram. It might be advisable to perform decontamination in the conditioning facility dedicated to operating waste in order to group all operations involving contaminating hazards in a single facility which is also equipped with a cementation unit for on-line treatment of decontamination effluents. By contrast, melting of metallic pieces and downstream metallurgical operations that cover very low-activity or inactive material could take place in a distinct facility.

4.2. OBJECTIVES

The minimum objectives to be reached are:

- the requirements set by safety authorities regarding safety and the acceptability of packages to be stored in final repositories; and

- the waste management allowing an exact inventory of radioactive products and thus the certification of the conditioned ware.

However, complementary objectives must be set by the plant owner in order to define the concept of a facility.

The most significant of these objectives are:

- plant capacity;
- volume reduction factor;
- reuse or not of some pieces; and

- the operating method that might favor automated and remotely controlled operations to decrease absorbed doses to operators.

The above criteria may lead to devising a facility built in two steps.

In the first step, minimum functions are implemented to treat the wastes as early as possible in compliance with safety and final disposal requirements. Neither the volume reduction factor nor the operations are optimized at this stage, but the structure and auxiliary functions of the facility account for future extensions.

In the second step, complementary equipment is installed to improve overall performance of the facility.

The functions to be implemented depend on the selected criteria.

4.3. CENTRALIZATION

Centralized facilities better utilize large equipment and auxiliary functions and thus drastically reduce waste treatment costs. Although solid waste transportation raises no major safety problems, it might be advisable to limit transport for both economic and public acceptance reasons.

Some preparatory functions can be performed at the production sites, for instance sorting or precompaction of burnable solids that involves inexpensive equipment and reduces the volume of the transported waste by a factor of three to four.

4.4. UTILIZATION OF FIXED OR MOBILE EQUIPMENT

The problem raised by radioactive waste transport can be bypassed using mobile facilities already operated on a commercial scale for liquid wastes (cementation, bituminization) and solids (supercompaction).

Their utilization presupposes that the equipment can be easily decontaminated and connected/disconnected to host facilities.

4.5. APPLICATION TO SOLID WASTES FROM THE MURMANSK REGION

The following scheme could be considered :

- Simultaneous treatment of operating and dismantling wastes. Melting metallic wastes from dismantling operations on an industrial scale is not recommended in the near future for the following reasons:

    - all high temperature processes are expensive;

- currently applicable regulations do not allow large scale release of slightly radioactive metals for public reuse; and
- melting could preferably be applied to non-contaminated waste.

- Construction of the facility in several steps to rapidly allow minimum treatment of current waste. In the first step, the following functions could be implemented:

  - reception of waste packages;
  - nuclear measurements for full characterization of the waste;
  - buffer storage;
  - decontamination using simple processes to remove non-fixed contaminants and facilitate subsequent cutting;
  - dismantling/cutting of bulky wastes mainly originating from dismantling operations; and
  - immobilization of the wastes in a cement matrix inside high integrity containers to obtain packages offering durable confinement for acceptance in final repositories.

- Other functions—such as incineration, more efficient waste decontamination, and metallic waste melting—will eventually be implemented later according to the evolution of regulatory and economic conditions.

CHAPTER 10
CATALYTIC EXTRACTION PROCESS (CEP) APPLICATIONS TO MIXED AND RADIOACTIVE WASTES AND WEAPONS COMPONENTS

Potential for Use in Northwestern Russia

MICHAEL LANDRUM, CHRISTOPHER A. HERBST, ERIC P.
LOEWEN, ANNA PROTOPAPAS, CLAIRE A. CHANENCHUK,
ESTHER W. WONG
Molten Metal Technology, Inc.
Washington D.C., United States

1. Abstract

Quantum-CEP™ (Q-CEP™) is an innovative and patented technology developed by Molten Metal Technology, Inc. (MMT) that can process radioactive and mixed waste streams to decontaminate and recover resources of commercial value while achieving significant volume reduction and radionuclide stabilization. MMT is currently in the process of commercially deploying Q-CEP™ in the government and commercial radioactive markets. To this end, MMT has formed two relationships with market leaders: Martin Marietta Corporation and the Scientific Ecology Group (SEG), a Westinghouse subsidiary. M4 Environmental, L.P., a fifty-fifty partnership between MMT and Martin Marietta Corporation, was formed to deploy Q-CEP™ applications in the United States Department of Energy (DOE) and Department of Defense (DoD). A Q-CEP™ demonstration facility for the processing of DOE mixed waste in a privatized mode is currently under design and construction. MMT has also joined forces with SEG to deploy Q-CEP™ systems for processing spent ion exchange resins from nuclear power plants worldwide. A United States facility is currently under construction and will be operational by the end of 1995.

This paper outlines a wide range of CEP/Q-CEP™'s processing results, including Resource Conservation and Recovery Act (RCRA) wastes, spent ion exchange resins, and contaminated scrap metal. These results highlight the technology's unique features in processing radioactive and mixed wastes:

1. *Waste Minimization Performance:* RCRA wastes are recycled into commercially valuable products: gases, ferroalloys, and ceramics. Processing of mixed waste streams leads to complete destruction of hazardous contaminants and the potential

E.J. Kirk (ed.), Decommissioned Submarines in the Russian Northwest, 113–125.

formation of one or more decontaminated product streams, such as fuel gas and/or metal alloys.

2. *Environmental Performance:* Processing of RCRA wastes leads to environmental performance surpassing regulatory standards. Destruction Removal Efficiencies (DREs) greater than or equal to 99.9999 percent have been consistently demonstrated. NO_x and SO_x were not detected to three parts per million (ppm) (analytically limited). Dioxins/furans were not detected to the targeted regulatory limit of 0.1 ng/Nm^3 toxicity equivalent (TEQ). All ceramic products passed Toxicity Characteristic Leaching Procedure (TCLP) tests.

3. *Controlled Radionuclide Partitioning:* Processing of contaminated scrap metal results in metal decontamination greater than 99 percent, with current experimental demonstrations limited by the lower detection limit (LDL) of the analytical equipment. Commercial processing of spent ion exchange resins will result in gas decontamination factors greater than 10^7.

4. *Final Form Stabilization:* Multiple high temperature glass compositions have been developed for Q-CEP™ applications. Scanning Electron Microscopy (SEM)/ Energy Dispersive X-ray Diffraction (EDX) results showed radionuclide incorporation into final form. Radionuclide incorporation into stable high chlorine content glass structures outperforming high-level nuclear waste reference glasses in Product Consistency Test (PCT) and TCLP durability testing has been proven.

5. *Volume Reduction:* Volume reduction from processing contaminated scrap metal is specific to individual scenarios and dependent on the initial form of the waste material. Processing of spent ion exchange resins results in volume reductions greater than 30 to one. Preliminary calculations indicate that volume reductions greater than 100 to one may be achievable while processing DOE mixed low-level waste streams (MLLW), such as organic sludges and combustible debris.

MMT operates a variety of experimental and demonstration-scale Catalytic Extraction Process (CEP) systems and has demonstrated long-term operability and reliability to commercial and government customers. Over 40 customer-established performance-based criteria covering environmental performance, product quality, and operability were surpassed with third-party oversight and validation by a major German chemical manufacturer. An on-stream factor greater than 90 percent was achieved.

2. Q-CEP™ Research, Demonstration, and Commercial Facilities

Catalytic Extraction Process (CEP) is an innovative and proprietary technology that allows waste material of a wide range of composition (organic, organometallic, and inorganic) to be recycled into products of commercial value, such as synthesis gas, hydrogen chloride, metal alloys, and specialty ceramics. At the core of CEP is a liquid metal bath which acts as a catalyst and solvent in the dissociation of the feed and the synthesis of products. Upon introduction to the bath, feeds dissociate into their constituent elements. Addition of co-reactants enables reformation and partitioning of desired products. The catalytic and solvation effects of CEP, together with the thermodynamically controlled reaction pathways, allow the technology to achieve superior environmental and waste minimization performance.

Q-CEP™ is the application of the CEP technology to process radioactive and mixed waste streams. Targeted radionuclide partitioning leads to decontamination and recycling of a large portion of the waste components to commercial products (e.g. synthesis gas) with high resultant volume reduction through radionuclide concentration into a stable condensed phase.

MMT has engaged in technical development of Q-CEP™ and is currently in the process of commercially deploying the technology to the government and the commercial radioactive market. MMT has formed two significant partnerships to enable quick deployment of Q-CEP™:

1. A Martin Marietta Corporation/MMT limited partnership, M4 Environmental, L.P., is designing and constructing a demonstration unit for application to DOE wastes; and

2. MMT and the Scientific Ecology Group (SEG), a Westinghouse subsidiary, are jointly constructing a Q-CEP™ facility to process radioactively contaminated spent ion exchange resins.

A variety of experimental and demonstration-scale CEP and Q-CEP™ units have been operated for both research and customer needs. MMT's Recycling-Research & Demonstration Facility in Fall River, Massachusetts, has been the primary site for technology development and customer demonstrations. Fall River houses four bench-scale units, five pilot-scale units, seven physical models, and a commercial demonstration prototype unit. The facility is fully permitted by the Commonwealth of Massachusetts for recycling demonstrations using hazardous and non-hazardous materials as CEP feeds. At Fall River, MMT has carried out a range of demonstrations on RCRA materials and has received recycling certifications from the Massachusetts Department of Environmental Protection for the processing of RCRA-listed and characteristic feeds. MMT has also carried out radioactive surrogate and common isotope testing at the Fall River facility. Testing on low-level radioactive materials, and particularly spent ion exchange resins, has taken place at MMT's demonstration facility in Oak Ridge, Tennessee. MMT is expanding its "hot" testing capabilities by deploying with M4 demonstration-scale Q-CEP™ units to process low-level waste and mixed low-level waste.

Technical development and commercialization of Q-CEP™ was partly sponsored by DOE. From 1993 to date, MMT has been awarded a total of $19 million under a Planned Research and Development Announcement (PRDA).[7] Contract-sponsored activities focused on demonstration of Q-CEP™ chemistry, recovery potential and final form stability, development of a conceptual engineering design, and application of Q-CEP™ to contaminated scrap metal and mixed low-level waste. Under the agreement with SEG, MMT is constructing a Q-CEP™ commercial facility at Oak Ridge for processing radioactively contaminated ion exchange resins from nuclear power plants. The Q-CEP™ system, designed for high volume reductions (greater than 30 to one), was targeted for operation in 1995.

An extensive demonstration run was successfully completed for the recycling of biosolids waste provided to MMT by a major German chemical manufacturer. The run demonstrated the long-term operability of the CEP system with respect to on-stream factor and steady-state operation. This prototype campaign also demonstrated CEP's utility for on-line recycling of a commercial-grade synthesis gas. A number of operational performance criteria for on-stream factors, steady-state operation, and product quality and consistency were established jointly by the customer and MMT. Based upon on-site third party customer validation, more than 40 operational criteria were met and surpassed (See Table 1). Steady-state operation was successfully demonstrated and product quality and consistency met the needs for on-line recycling. An on-stream factor greater than 90 percent was achieved [6]

TABLE 1. Steady-State CEP Operability Demonstration—Run Performance Criteria

Measurement	Method of Measurement	Target	Actual
Process gas flow	Ar tie balance	±50 SCFM*	±20 SCFM* at steady feeds
Process gas comp.—CO	Mass spectrometer	±10 vol%	±2 vol%. at steady feeds
Process gas comp.—H_2	Mass spectrometer	±10 vol%	±2 vol%. at steady feeds
Process gas comp.—THC	THC analyzer	< 200 ppm	< 200 ppm
Bath carbon	LECO carbon analyzer	±1 wt%	±0.8 wt% over entire run
Bath temperature	Thermocouples	±200°F	±100°F over entire run

* SCFM = Standard Cubic Feet Per Meter

3. Q-CEP™ Processing of RCRA Waste

CEP has been demonstrated on a range of RCRA listed wastes as well as characteristic and RCRA-like surrogate material (See Table 2), including high-molecular weight aromatics, chlorinated organics, organically-bound nitrogen species (isocyanates), plastics, and organometallics. Table 2 data are from waste processing, using the Fall River demonstration unit. The technology's environmental performance has been demonstrated during actual waste processing to meet and surpass current and proposed regulatory standards. Specifically, DREs greater than 99.9999 percent were achieved for principle organic hazardous constituents (POHCs). NO_x and SO_x were not detected in the product gases to detection limits of three ppm. Condensed phase non-leachable products, both ceramic and metal phases, of marketable composition were generated.

TABLE 2. Major Feeds Processed at the Demonstration Prototype

Feed	Waste Minimization Performance			Environmental Performance		
	Product Recovery	% of Feed Recycled	Residual	DRE	TCLP	Off-Gas
Representative Surrogate Feeds						
Polystyrene/Graphite	Syngas	99% to syngas	Dust Negligible	≥99.9999%	N/A	PICs< LDL[1] of 1 ppm CO_2<1% NO_x, SO_x < LDL of 100 ppm
Chlorotoluene/ Heavy Organics	Syngas Ceramic	87% to syngas 12% to ceramic	Dust Negligible	≥99.9999%	Passes TCLP	PICs< LDL of 1 ppm CO_2<1% NO, SO_x < LDL of 100 ppm
Dimethyl Acetamide/ Heavy Organics	Syngas Nitrogen	96% to syngas 3% to nitrogen	Dust Negligible	≥99.9999% (based on THC)	N/A	PICs< LDL of 1 ppm CO_2<1% NO, SO_2 < LDL of 100 ppm[2] NO_x SO_x < LDL of 1 ppm[3]
Representative Hazardous Waste Feeds						
Industrial Biosolid Waste	Syngas Nitrogen Ceramic Ferroalloy	70% to syngas 8% to nitrogen 20% to ceramic 1% to ferroalloy	Dust Negligible	≥99.9999% (based on THC)	Passes TCLP	PICs< LDL of 1 ppm CO_2<1% NO_x, SO_x< LDL of 3 ppm Trace<0.1 ng/Nm^3 TEQ[7]
Surplus Metal/ Weapon Componentry	Syngas Ceramic Ferroalloy	25% to syngas 8% to ceramic 63% to ferroalloy	Dust Negligible	N/A	Passes TCLP	PICs< LDL of 1 ppm CO_2<1% NO, SO_2 < LDL of 100 ppm[2] Trace<0.1 ng/Nm^3 TEQ[7]
K019/K020[4]/ Chlorobenzene/Fuel oil	Syngas	76% to syngas 23% to ceramic	Dust Negligible	≥99.9999%	Passes TCLP	PICs< LDL of 1 ppm; CO_2<1% NO, SO_x < LDL of 100 ppm[2]
K027[5]	Syngas Ferroalloy (Fe-Ni)	93% to syngas 5% to nitrogen <1% to ceramic <1% to ferroalloy	Dust Negligible	≥99.9999%	Passes TCLP	PICs< LDL of 1 ppm CO_2<1% NO, SO_x < LDL of 100 ppm[2]
F024[6]/Fuel oil/ Chlorotoluene	Syngas HCl gas	82% to syngas 13% to HCl gas <1% to ceramic	Dust Negligible	≥99.9999%	Passes TCLP	PICs< LDL of 1 ppm CO_2<1% NO, SO_x < LDL of 100 ppm[2] Trace<0.1 ng/Nm^3 TEQ[7]

[1] LDL = Lower Detection Limit
[2] As measured by on-line mass spectrometer. Lower detection limit to 1 ppm was confirmed as in footnote 3.
[3] As measured by third party analytical equipment placed on-line.
[4] EPA listed hazardous waste stream: ethylene dichloride/vinyl chloride heavy ends.
[5] EPA listed hazardous waste stream: toluene diisocyanate distillation residues.
[6] EPA listed hazardous waste stream: chlorinated aliphatic hydrocarbons.
[7] Dioxins and furans nondetectable to the targeted regulatory limit of 0.1 ng/Nm^3 TEQ.

Dioxins were not detected to the targeted regulatory limit of 0.1 ng 2,3,7,8 tetrachloro-dibenzo-para-dioxin (TCDD) TEQ/Nm^3.

Currently, the effectiveness of hazardous waste treatment technologies is regulated by the United States Environmental Protection Agency (EPA) by measuring the concentration of hazardous materials in the after-process wastewater and non-wastewater. All processes must ensure that organic constituent concentrations in individual wastewater and non-wastewater streams are lower than the regulated limits set by EPA. CEP demonstrations surpass the current limits even when compared with the Best Demonstrated Available Technology (BDAT). CEP was recently approved by EPA (on 24 October 1994) as a non-combustion technical equivalent (BDAT) for eight RCRA-listed isocyanate waste codes (K027) for which incineration had previously been mandated as the commercially available BDAT technology.

TABLE 3. CEP Performance in BDAT Constituents Conversion (F024 Processing)

Constituents	Feed (mg/l)	Non-Wastewater Regulations			Wastewater Regulations	
		Ceramic[3] (mg/kg)	Metal[3] (mg/kg)	Non-Wastewater (mg/kg)	Scrubber Water	Wastewater (mg/l)
1,1,2-Trichloroethane	18,000	ND[1] (0.0075)[2]	ND (0.0075)	6.0	ND (0.0015)	0.054
Tetrachloroethene	9,700	ND (0.0075)	ND (0.0075)	6.0	ND (0.0015)	0.056
Chlorobenzene	12,000	ND (0.018)	ND (0.018)	6.0	ND (0.0035)	0.057
1,2-Dichloroethane	6,800	ND (0.0075)	ND (0.0075)	6.0	ND (0.0015)	0.21
1,1,2,2-Tetrachloroethane	11,000	ND (0.005)	ND (0.005)	6.0	ND (0.001)	0.057
Trichloroethene	24,000	ND (0.005)	ND (0.005)	6.0	ND (0.001)	0.054
Xylenes	610	ND (0.005)	ND (0.005)	3.0	ND (0.001)	0.32
1,1,1,2-Tetrachloroethane	1,000	ND (0.025)	ND (0.025)	6.0	ND (0.001)	0.057
Hexachlorobutadiene	3,600	ND (0.025)	ND (0.025)	5.6	ND (0.001)	0.055
Naphthalene	900	ND (0.025)	ND (0.025)	5.6	ND (0.001)	0.059

[1]ND = Not Detected. [2](n) where n is the lowest detection limit. [3]In a commercial facility processing F024, the metal and ceramic would be products, not wastes subject to LDR standards. These data are provided to demonstrate CEP conversion of BDAT constituents.

The environmental performance of CEP has also been demonstrated on chlorinated waste streams containing some of the most difficult to destroy hazardous constituents.[1] The results of demonstration-scale processing of RCRA-listed waste F024 (chlorinated aliphatics) indicated that hazardous organic constituents in the feed were not detected in ceramic, metal, and scrubber water, which surpassed BDAT standards for all effluent streams (See Table 3). Destruction Removal Efficiency (DRE) on multiple organic hazardous constituents exceeded 99.9999 percent, surpassing current regulations mandating DREs greater than 99.99 percent (See Table 4). Trace constituents were not detected to the targeted regulatory limit of 0.1 ng/Nm^3 TEQ (See Table 5).

TABLE 4. Destruction Removal Efficiency (DRE) for Listed Hazardous Constituents (F024 Processing)

Feed Constituent	Feed Concentration (mg/l)	Off-gas Concentration (ppm)	DRE
1,1,2-Trichloroethane	18,000	ND[1] (0.00037)[2]	≥99.9999%
Chlorobenzene	12,000	ND (0.00037)	≥99.9999%
1,2-Dichloroethane	6,800	ND (0.00044)	≥99.9999%
1,1,2,2-Tetrachloroethane	11,000	ND (0.00029)	≥99.9999%
Trichloroethene	24,000	ND (0.00038)	≥99.9999%

[1]ND=Not Detected [2](n) where n is the lowest detection limit.

TABLE 5. Trace Component Off-Gas Concentrations During CEP of Chlorinated Wastes

Description:	Aromatics (20% PVC)	Biosolids (2% PVC)	Biosolids (2% PVC)	F024	F024
Analytes (ng)					
Total TCDD (tetrachlorodibenzo-para-dioxin):	ND	ND	ND	ND	ND
Total PeCDD (pentachlorodibenzo-para-dioxin):	ND	ND	ND	ND	ND
Total HxCDD (hexachlorodibenzo-para-dioxin):	ND	ND	ND	ND	ND
Total HpCDD (heptachlorodibenzo-para-dioxin):	ND	ND	ND	ND	ND
Total OCDD (octachlorodibenzo-para-dioxin):	ND	ND	ND	ND	ND
Total TCDF (tetrachlorodibenzofuran):	ND	ND	ND	ND	ND
Total PeCDF (pentachlorodibenzofuran):	ND	ND	ND	ND	ND
Total HxCDF (hexacholorodibenzofuran):	ND	ND	ND	ND	ND
Total HpCDF (heptachlorodibenzofuran):	ND	ND	ND	ND	ND
Total OCDF (octachlorodibenzofuran):	ND	ND	ND	ND	ND
2,3,7,8-TCDD toxicity equivalents (ng/Nm^3) EPA-1989[1]	ND	ND	ND	ND	ND

ND = Not Detected to targeted regulatory standard of 0.1 ng/Nm^3 TEQ.
[1]Determined as the sum of each individual analyte concentration multiplied by its corresponding toxic equivalent factor.

4. Q-CEP™ Application to Spent Ion Exchange Resins

Nuclear power plants in the United States generate over 200,000 cubic feet of spent ion exchange resin per year. Ion exchange resins are widely used in nuclear power plants for water purification. The resin waste stream represents the largest waste stream in both volume and activity for nuclear power plants. Regeneration of contaminated ion exchange resin is not economically feasible due to the large volume of contaminated liquids generated and other market barriers. Disposal has been hampered by the recent closure of the Barnwell, South Carolina radioactive waste disposal facility to many of the nuclear power plants. A technology providing volume reduction and stabilization offers significant economic and safety benefits to nuclear power plants.

4.1. SURROGATE TESTING

Prior to hot testing, MMT performed demonstration runs on nonradioactive resin at the Fall River demonstration unit to study CEP efficiency in organic resin conversion. Gas-phase detection of the resin (polystyrene-divinylbenzene polymer) is limited by sampling protocol; hence, a modified method was used to calculate DRE. Using styrene as a decomposition product, the calculated DRE was greater than 99.9999 percent (below the LDL of 1.2 parts per billion [ppb]). As a confirmation, the destruction efficiency for the resin was also calculated by comparing the number of moles of benzene rings in the solid feed with the number of moles of benzene rings in the gas phase. This analysis also resulted in a DRE greater than 99.9999 percent.

4.2. BENCH-SCALE TESTING

A range of "hot" tests using commercial spent ion exchange resin have been performed at SEG's facility in Oak Ridge using bench-scale Q-CEP™ systems. A representative total curie balance is shown in Table 6 for the gamma-emitting radionuclides in the resin supplied by SEG. A radionuclide closure of 105±7 percent was measured.

TABLE 6. Curie Balance Closure From Radioactive Ion Exchange Resin Processing

Resin Composition	Injected Weight: 64.5g			
Nuclide	μCi Injected	Feed %	μCi Measured	% Recovered
Co-58	9.37×10^{-03}	0.44	1.01×10^{-02}	107.69
Co-60	7.10×10^{-01}	32.97	8.15×10^{-01}	114.86
Cs-134	1.30×10^{-01}	6.05	1.34×10^{-01}	102.68
Cs-137	4.83×10^{-01}	22.44	4.67×10^{-01}	96.68
Mn-54	1.62×10^{-01}	7.54	1.75×10^{-01}	108.10
Zn-65	6.58×10^{-01}	30.57	6.56×10^{-01}	97.78
Total Curies	2.15	100.00	2.22	104.81

Table 7. System Decontamination Factors

Nuclide	Decontamination Factors	Primary Partitioning Phase
Co-60	≥ 65,000	metal bath
Cs-134	≥ 82,000	volatile radionuclide trap
Cs-137	≥ 132,000	volatile radionuclide trap
Mn-54	≥ 125,000	metal bath
Zn-65	≥ 161,000	volatile radionuclide trap

Note: Decontamination factors are analytically limited, as the radionuclides are below the equipment detection levels.

The bench-scale testing has demonstrated decontamination factors leaving the reactor system greater than or equal to 10^4 (limited by analytical). The commercial facility is designed for decontamination factors greater than 10^7. Table 7 shows representative decontamination factors achieved in the bench-scale tests exiting the reactor system but excluding the gas handling train (i.e. HEPA filter). Decontamination factor is defined as DF = Activity In/Activity Out.

4.3. COMMERCIAL FACILITY

The Q-CEP™- ion exchange resin (IER) commercial facility, jointly operated by SEG and MMT, will be capable of processing up to 400 High Integrity Containers (HICs) or 80,000 cubic feet of IER per year. The facility will operate two Q-CEP™ systems in a batch mode, which will prevent commingling of customer waste. A customer batch of one or more HICs will be processed, and a stable final form contained in a steel container will be returned to the customer. Volume reduction is determined by operating parameters such as customer batch size, resin composition, and curie content. The design specifies that volume reductions will be at least 30 to one, while ongoing optimization of operating parameters is expected to lead to higher volume reduction ratios.

5. Q-CEP™ Application to Contaminated Scrap Metal

Previously published studies have demonstrated the ability to partition radioactive components, such as uranium and plutonium, from the metal phase into a vitreous phase via melt refining.[4,5,8,9] Residual concentrations ranging from 0.05 to two ppm were achieved using diffusion of oxidizing, vitreous-forming agents to partition the radioactive components. Successful partitioning to less than ten nCi/g levels (approximately less than 0.1 ppm) has been demonstrated for uranium- and plutonium-contaminated metals. Q-CEP™ offers the potential for superior performance, as the techniques involved in melt refining are completely incorporated and enhanced in Q-CEP™ technology. Specifically, Q-CEP™ incorporates active radionuclide partitioning through select coreactant additions (e.g. oxygen) and enhanced mass transfer (e.g. convection), while melt refining is based upon "passive" diffusion-based partitioning. MMT has demonstrated the capabilities of Q-CEP™ to process contaminated scrap metal under the DOE-sponsored development programs.

5.1. BENCH-/PILOT-SCALE TESTING ON SURROGATES

Q-CEP™ processing of radioactive wastes was first demonstrated using surrogates at the Fall River facility. Hafnium and cerium were chosen as the surrogates of radionuclides uranium and plutonium based on similarities in oxidation free energy, density, and method of incorporation into glass.[3] Theoretically, these elements will be oxidized and captured in the ceramic phase. Bench-scale tests were carried out varying experimental parameters including metal charge (nickel and iron), glass composition (a matrix of calcium alumino/borosilicates), temperature, and bath carbon level. Initial contamination levels ranged from 350 ppm to 900 ppm. Table 8 shows representative decontamination factors from the processed metal samples. The results indicate successful removal of hafnium and cerium from the "contaminated" metal phase, with demonstrated decontamination factors greater than 99 percent (LDL limited). Overall mass balance closure greater than 97 percent has been demonstrated.

TABLE 8. NAA Results for Hf/Ce Partitioning from the Metal Phase to the Ceramic

Metal Charge	Glass Composition	Contaminant	Metal [1,2,3,4] Decontamination (%)
Iron	Aluminosilicate	Hafnium	≥ 99.96
Iron	Aluminosilicate	Hafnium	≥ 99.80
Nickel	Aluminosilicate	Hafnium	≥ 99.90
Nickel	Aluminosilicate	Cerium	≥ 99.14

[1]Based on the concentration of Hf or Ce detected by Neutron Activation Analysis. LDL in F=1 ppm; LDL in Ni=0.2 ppm.
[2]Amount of Hf or Ce removed. Calculation based upon individual data points.
[3]Removal of surrogate from the metal phase, decontamination = (Hf initial – Hf final) / Hf initial * 100%.
[4]Limited by the lowest detection limit of the equipment.

5.2. BENCH-SCALE TESTING OF CONTAMINATED SCRAP METAL

MMT continued Q-CEP™ experiments in the bench-scale units at Oak Ridge using depleted uranium (U) and uranium/cerium (Ce) mixtures. The experimental program included optimization of the ceramic phase capture of these species. The primary experimental design parameters evaluated were glass composition and initial form/loading of the contaminant (See Table 9).

TABLE 9. Uranium Partitioning Study

Experimental Parameter	Levels Studied
Glass Composition	2: Calcium Aluminosilicate and Borosilicate
Initial Metal Composition	4: Iron (Fe) with Uranium Loadings of 0.1, 0.5, and 1.0 wt% and Cerium Loadings of 0.1 wt%
Contaminants	2: Depleted Uranium (from 1-5 wt% in final glass) and Cerium. Uranium added as both U metal and as UO_2, cerium added as CeO_2.
Temperature	1: 1500–1575°C
Carbon Level	1: 3±0.5 wt%

Samples were taken both from the metal surface and throughout the interior and analyzed for uranium and cerium content. The ceramic phase, reactor crucible, headspace insulation, and gas handling train were also subjected to radiochemical analyses. Greater than 95 percent curie closure was achieved.

Scanning electron microscopy (SEM) and Energy Dispersive X-ray Diffraction (EDX) were performed on selected glass samples to investigate the nature of incorporation into the ceramic phase. Results indicated that both uranium and cerium were partitioned to the targeted phase (glass). Figure 1 shows representative results demonstrating the partitioning and capture of uranium in the vitreous phase. These EDX results originated from a Q-CEP™ experiment in which the initial iron metal charge contained 1000 ppm uranium. This EDX clearly shows that uranium is present in the glass phase along with the calcium, aluminum and silicon (peaks have been labeled). Uranium decontamination is further quantified by neutron activation analysis (NAA) studies. In an iron-aluminosilicate-uranium system, decontamination of uranium was found to be greater than or equal to 99.77 percent (LDL limited), which demonstrated consistent findings with surrogate tests. Decontamination measurement was limited by the analytical LDL, which was on the order of 0.1 ppm (dependent upon the contaminant and metal). Thermodynamic calculations indicate that the residual contaminant concentration in the metal should be several orders of magnitude lower than the detection limit of NAA. To verify the stability of the glass formed, PCT and TCLP tests were performed on a series of synthesized glasses that contained various amounts of RCRA metals and uranium. All glasses were shown to be significantly more durable (i.e. less leachable) than high-level nuclear waste reference glasses.

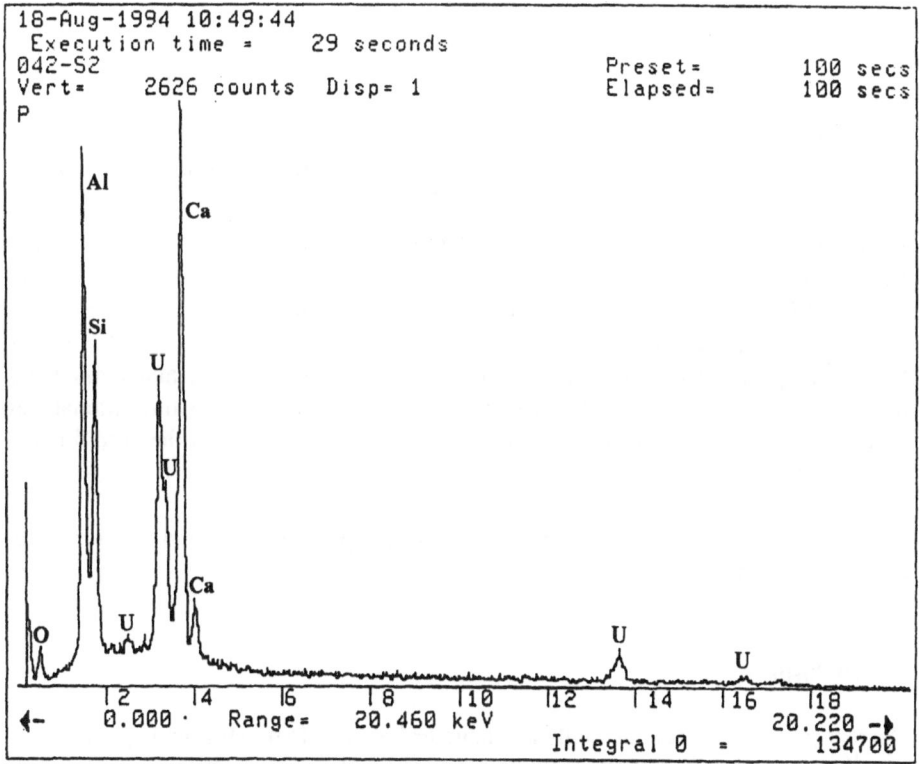

Figure 1. EDX Spectrum of Waste Glass Showing Presence of Uranium

6. Q-CEP™ Application to Low-Level Mixed Waste

6.1. VOLUME REDUCTION

Volume reduction calculations have been carried out based on Q-CEP™ of DOE mixed low-level waste (MLLW) streams (See Table 10). Major waste streams include soils, inorganic sludges, organic sludges, combustible debris, and inorganic debris. The ceramic phase, which captures the radionuclide contaminants, is expected to be the only solid phase which may require disposal. In the absence of a *de minimis* standard, decontaminated metal has restricted reuse applications within the DOE complex. Given an achievable *de minimis* standard, free release into the commercial sector would also be allowed. The synthesis gas product, which captures essentially all carbon and hydrogen in the feed, may be used within the processing system for energy generation. The volume reduction varies from two to one, to greater than 100 to one, depending on the initial form and composition of the feed material.

TABLE 10: Expected Volume Reduction for MLLW Processing Via Q-CEP™

	Initial Mass (kg)	Initial Bulk Density (kg/m³)	Initial Volume (m³)	Final Ceramic Mass (kg)	Final Ceramic Volume (m³)	Volume Reduction
Soil	1000	1785	0.56	769	0.31	2:1
Inorganic Sludge	1000	1300	0.77	209	0.08	10:1
Organic Sludge	1000	900	1.11	5	0.002	>100:1
Combustible Debris	1000	135	7.4	78	0.03	>100:1
Inorganic Debris	1000	1700	0.59	750	0.3	2:1

A recent DOE-sponsored evaluation of Integrated Thermal Treatment Systems [2] for application to DOE mixed wastes showed that a Q-CEP™ system would have the lowest total life cycle costs (TLCC) compared to eighteen alternative systems, including incineration, vitrification, and plasma treatment. Q-CEP™ offers a one-step process that simultaneously achieves organics conversion, metal recovery, and radionuclide immobilization. Q-CEP™'s broad feed acceptability and robust nature dramatically reduce the feed pretreatment and characterization requirements and the need for post-processing stabilization.

This streamlined system design lowers capital and operating costs and enhances overall system volume reductions.

7. Conclusions

Currently, no single commercially available technology dominates the huge market of low-level and mixed radioactive waste generated and stored in the United States and abroad. Q-CEP™ can be applied to a wide range of low-level radioactive and mixed waste feeds, effectively partitioning and stabilizing radionuclides while destroying hazardous constituents. Furthermore, Q-CEP™ may be able to reduce radioactive levels in recovered materials to below background levels, which may facilitate the establishment of *de minimis* standards. Q-CEP™ applications to DOE and commercial radioactive waste problems have been demonstrated and outlined in this paper. Current commercialization programs will deploy the technology to both the government and commercial sectors. A Q-CEP™ facility accepting spent ion exchange resin from nuclear power plants is targeted for operation in 1995. A mixed waste demonstration facility to process DOE wastes in a privatized mode is also under design and construction for 1995.

References

1. Chanenchuk, Claire A.; Protopapas, Anna; and Alexopoulos, George. (September 1994) *Catalytic Extraction Process Application To Chlorinated Waste Streams,* presented at the I&EC Special Symposium, American Chemical Society.

2. Feizollahi, Fred, and Quapp, William J. (November 1994) *Integrated Thermal Treatment System Study Phase 2 Results,* Idaho National Engineering Laboratory, DOE Contract No. DE-AC07-94ID13223.

3. Herbst, Christopher A., Loewen, Eric P., Nagel, Christopher J., and Protopapas, Anna. (April 1994) *Quantum-Catalytic Extraction Process Application to Mixed Waste Processing,* presented at the Nuclear and Hazardous Waste Management International Topical Meeting, American Nuclear Society.

4. Heshmatpour, G.L., Copeland, G.L., Heestand, R.L. (December 1981) *Decontamination of Transuranic Waste Metal by Melt Refining,* Oak Ridge National Laboratory, DOE Contract No. W-7405-eng-26).

5. Heshmatpour, B. and Copeland, G.L. (January 1981) *The Effects of Slag Composition and Process Variables on Decontamination of Metallic Wastes by Melt Refining,* Oak Ridge National Laboratory, DOE Contract No. W-7405-eng-26.

6. Mather, Robert; Celanese, Hoechst; Steckler, David; Tanner, Al; and Daniel, Fluor. (March 1995) *Integrated Recycling of Industrial Waste Using Catalytic Extraction Processing at a Chemical Manufacturing Site,* presented at AIChE Spring National Meeting.

7. Molten Metal Technology, Inc. (December 1994) *Recycle of Contaminated Scrap Metal,* DOE Contract No. DE-AC21-93MIC30171.

8. Seitz, M.G., Gerding, T.J., and Steindler, M.J. (June 1979) *Decontamination of Metals Containing Plutonium and Americium,* Argonne National Laboratory, Document No. ANL-78-13.

9. Uda, T., Iba, H., and Tsuchiya, H. (April 1986) Decontamination of Uranium-Contaminated Mild Steel by Melt Refining, *Nuclear Technology,* 73, 109.

IV. RISK ASSESSMENT AND MONITORING TECHNIQUES

CHAPTER 11
ASSESSMENT OF RADIOACTIVE CONTAMINATION IN THE ARCTIC

Status Report from AMAP

PER STRAND
Norwegian Radiation Protection Authority
Oslo, Norway

1. Introduction

The Arctic Monitoring and Assessment Program (AMAP) is pursuing an ongoing evaluation of the nature and consequences of past and present radioactive contamination and risks posed by potential sources. This work will result in a report to the Ministers of Environment in the Arctic countries in 1997. The assessment of radioactivity is led by Russia and Norway. The drafting group currently comprises participants from Russia, Norway, Canada, Denmark, Finland, Germany and the United Kingdom. National experts from all AMAP countries are also participating in the process.

At the Ministerial Conference in Nuuk, Greenland, the Ministers of the Environment asked AMAP to establish a data center in cooperation with appropriate international and national agencies, "to establish databases of sources, type, and levels of radionuclide contamination of the atmospheric, the aquatic, and the terrestrial environments of the Arctic and the Northern areas." The AMAP Data Center commenced operation in 1995 and is making an important contribution to the assessment. The United States, Russia, and Norway have taken on special responsibility for creating and operating the Data Center, which is sponsored by Norway and the United States. The Center is also cooperating with international organizations such as the International Atomic Energy Agency and the European Union to ensure effective coordination and avoid unnecessary overlap with other projects. Some of the information presented in this report, which addresses current knowledge, is based on data collected and analyzed at the Data Center.

The expert group convened for the environmental radioactivity assessment has the following major objectives:

• to describe historical and current levels of radioactivity in the Arctic;

E.J. Kirk (ed.), Decommissioned Submarines in the Russian Northwest, 129–131.

- to provide an overview of the different sources of radioactivity present in the Arctic area;

- to compare the importance of different radioactive sources for current or potential radioactive contamination of the Arctic environment;

- to provide individual-related risk assessments pertinent to Arctic populations; and

- to provide a data center to collate information on radioactivity in the Arctic area.

2. Transport Mechanisms

Radionuclides may be injected directly into the Arctic, transported within the Arctic, or brought to the Arctic from other areas of the globe through physical mechanisms such as atmospheric dispersion and deposition, ocean currents, and river and ice movement.

Radionuclides injected into the atmosphere (e.g. from weapons testing) are deposited on the Earth's surface through wet and dry atmospheric fallout. These radionuclides can become incorporated into soils, sediments, and surface waters from which they can subsequently be redistributed by various mechanisms such as surface runoff, ocean currents, ice movement, and eolian transport. Radionuclides released directly into waters (e.g. releases from Western European reprocessing plants) can be transported by water and incorporated into distant sediments.

In the Arctic, there is less precipitation than at temperate latitudes and hence, the deposition of globally distributed radioactivity is relatively lower than in more southerly regions of the northern hemisphere. Within the world ocean circulation system, the Arctic Ocean plays a decisive role. The relatively rapid transport of ice within the system can have a significant impact on the mobilization and subsequent relocation of sediment-bound radionuclides to other areas of the North Atlantic Ocean.

3. Environmental Impact of Past and Present Contamination

Maximum levels of man-made radionuclides in the Arctic environment occurred during the 1960s as a consequence of atmospheric nuclear weapons testing. Since the Interim Nuclear Test Ban Treaty of 1962, fallout radionuclide concentrations have declined substantially. Liquid effluent releases from nuclear fuel reprocessing operations in Western Europe have also contributed to radionuclides in the Arctic. Release rates from these plants have declined substantially since the mid-1970s. The combination of these events has resulted in a less pronounced decline in the concentrations of radionuclides in marine, as opposed to terrestrial, ecosystems. The third major source of radionuclides in the Arctic was the Chernobyl accident. Contributions from this event were more pronounced in the Eurasian Arctic than elsewhere.

The Radioactivity Assessment Group has reached the following preliminary conclusions as a contribution to the process:

1. Man-made radionuclides in the Arctic environment arise from three main sources: global fallout from atmospheric nuclear weapons tests, Western European reprocessing plants, and Chernobyl fallout.

2. Artificial radionuclide concentrations in the Arctic environment reached maximum levels in the mid-1960s and have subsequently declined throughout most Arctic areas.

3. Current and historical exposures from man-made radioactivity are lower for general Arctic populations than those arising from natural radioactivity.

4. Some indigenous peoples have been, and are, exposed to higher doses from man-made radioactivity than general Arctic populations and are accordingly at an increased health risk. This situation arises because these populations consume large quantities of wild food products—such as reindeer meat, freshwater fish, game, mushrooms, and berries—which have higher contamination levels of man-made radionuclides than agricultural food products.

5. The marine exposure pathways are less important than the terrestrial and freshwater pathways in contributing to human radionuclide exposure. Increased attention should be paid to terrestrial pathways in continuing assessments of radiation exposure in Arctic regions, particularly with regard to atmospheric releases of radioactivity.

6. Radioactive materials dumped in the Arctic marine environment do not currently pose a significant additional radiation risk.

7. In parts of the Arctic, large inventories of radioactive materials remain and pose potential risks. There is a need to demonstrate that operational procedures for decommissioning and maintenance of nuclear materials in the Arctic are carried out in a responsible and low-risk manner.

8. There is a considerable lack of information about the likelihood of accidents arising from such sources. It is important that information be provided by national authorities regarding risk and impact assessments. This would facilitate international evaluations of the adequacy of measures taken to reduce both the probability and severity of accidents.

CHAPTER 12
SPENT NUCLEAR FUEL ISSUES ON THE KOLA PENINSULA

KNUT GUSSGARD
E-Plan Knut Gussgard
Oslo, Norway

1. Abstract

This presentation addresses spent nuclear fuel on the Kola peninsula, where large quantities of spent fuel, mostly from nuclear submarines, have been stored secretly for decades. Since 1993, Russian authorities have gradually released information about storage sites and fuel conditions and quantities. Spent nuclear fuel is stored on land near military bases in the Litsa fjord at Andreev Bay and in Gremikha, as well as in floating facilities at the icebreaker base in Murmansk. Nuclear fuel still remains in some 60 submarines that have been taken out of service and are awaiting dismantlement. Norway has various nuclear programs which aim to increase nuclear safety at the Kola peninsula. These include one nuclear power plant, nuclear-operated submarines and icebreakers, and nuclear weapons materials.

2. Introduction

The Norwegian media and world media began to write about nuclear matters in Russia after the Chernobyl accident and accidents with nuclear submarines. The sinking of *Komsomolets* in 1989, for instance, attracted much attention in the media. Figure 1 is an image of *Komsomolets,* which now rests on the bottom of the Atlantic Ocean. In the same year, some episodes with submarines also occurred along the Norwegian coast. The picture presented in Figure 2 appeared in color on the front pages of Norwegian newspapers in 1989. It depicts an Echo II submarine, which was supposedly being towed back to base in Murmansk. This image is not of a towing. It shows a rubber hose transferring fresh water to cool the reactor on board the submarine, which had suffered a loss-of-coolant accident, resulting in a release of radioactivity inside the ship and large radiation doses to some of the crew. This incident attracted much attention because very little was known about Russian nuclear activities at the time.

There was, however, much discussion in the media, and the Russian White Book (the Yablokov report) became available in 1993, discussing the facts and problems related

E.J. Kirk (ed.), Decommissioned Submarines in the Russian Northwest, 133–140.

Figure 1. *Komsomolets*

Figure 2. The Echo II incident. Fresh Water for Emergency Cooling is Transferred Via a Rubber Hose from the Assisting Ship, *Konstatin Yuon*

to nuclear activities in the Kola area and in northwestern Russia. Among other things, the White Book described the storage of spent nuclear fuel on land. It stated that approximately 20,000 fuel assemblies from submarine reactors were stored on land at Andreev Bay while an estimated 4,500 spent fuel assemblies were stored on ships (the *Lepse*, the *Lotta*, and the *Imandra*) belonging to Atomflot, the icebreaker fleet in Murmansk. These figures correspond to roughly 100 reactor cores at Andreev Bay and some 20 reactor cores in Murmansk.

3. Dumped Fuel

Dumped reactors with spent fuel also attracted much attention in the press, radio, and television all over the world. News of them led Norway to develop a cooperation with the Russians to study the dumped reactors and to get better descriptions of them. The Norwegians proposed a cruise to visit these dumping sites to have a look at them and to take measurements around the objects in question. A team of scientists from Norway, Europe, and Russia went to these places with a Russian research vessel. Norwegian authorities initiated the establishment of formal and informal groups of scientists, some of which were international, to study observations from the cruise and new information released by Russian authorities. In the course of these investigations, the scientists learned a lot about the dumped reactors. Gradually, they also learned more about the spent fuel stored on ships in Murmansk, but very little detailed information was provided concerning Russian ship reactors, in general.

4. Fuel Stored on Land

4.1. SPENT FUEL

In 1996, however, the spent fuel in Andreev Bay was described in a Russian statement called Statement Number One, which was presented at a meeting of the North Atlantic Cooperation Committee (NACC), a NATO committee in which Russian and Eastern European countries also take part. The meeting was held in Rome earlier this year, and it was reconfirmed there that more than 20,000 fuel assemblies are stored at Andreev Bay. Some information about the fuel stored in Gremikha was also provided from the official Russian side at that time, which was, to my knowledge, the first time that an official Russian statement about the spent fuel in Gremikha was issued. Some of the fuel there is damaged and in a state similar to that of the fuel in the *Lepse*. There is also fuel from Russian metal cooled reactors in Gremikha. Many of the metal cooled reactors had serious accidents. Table 1 contains confirmed Russian information on nuclear fuel at Gremikha.

As others have already explained, most of the radioactivity in spent nuclear fuel and radioactive waste at the Kola peninsula is in the fuel itself. There is 1,000 times more radioactivity in the fuel than in the intermediate level solid waste in storage, and another 1,000 times more radioactivity in the fuel than in liquid radioactive waste, which means that liquid low-level radioactive waste contains one millionth of the radioactivity that spent fuel contains.

TABLE 1. Nuclear Fuel at the Technical Base Gremikha.

Basin No. 2, Spent Fuel Storage	95 spent fuel assemblies, damaged, not accepted for reprocessing
Open Air Storage	116 containers, type 6 (812 fuel assemblies), removed in 1960s, not accepted for reprocessing for technical reasons
Facility No. IV	5 complete reactor cores for submarine types 705, 705K, 745. No technology for reprocessing, nor equipment for transport.*
Open Air Storage	11 containers, type 11, reprocessing possible.

*This is fuel for metal cooled reactors.

4.2. NEW FUEL

This sums up and describes the risks associated with spent fuel. However, there is also concern from the Norwegian side about new fuel. This concern is related to the risk of proliferation of highly enriched uranium. Modern ship reactor fuel in Russia is 90 percent enriched uranium, often alloyed to other metals, which can easily be extracted by any chemist. The fuel assemblies in one nuclear reactor core, taken together, contain some 100 or 200 kilograms of highly enriched uranium. That is a sufficient amount of highly enriched uranium for approximately ten nuclear explosion devices. Such a reactor core, which is only one cubic meter in volume, can easily fit into a big van or a fishing boat, so the risk of theft and other means of proliferation of that material is obvious, which means the material must be guarded. This situation presents a real risk associated with the storage of new, fresh fuel for nuclear reactors, rendering physical protection of highly enriched uranium fuel, which can easily be transformed into a nuclear explosive, very important.

5. Risk Assessment

5.1. POTENTIAL RISKS

5.1.1. Criticality

As mentioned in previous chapters, a criticality accident—an accident with an uncontrolled chain reaction in the material—cannot be ruled out in spent nuclear fuel. Criticality accidents with fuel from nuclear submarines have occurred on board ships during fuel handling operations. There is also a possibility of criticality in stored fuel. Fires, explosions, or meltdown of the fuel may occur. In these cases, the possibility of a large release exists. Ølgaard (Chapter 2) pointed out that the reactor core from a submarine is much smaller than a reactor core in a nuclear power station. It is true that it is smaller, but it is not much smaller. The size of Russian reactors with respect to power in modern, new, third- and fourth-generation reactors is up to 200 megawatts, perhaps even greater in a few cases. That is as much as 15 percent, for instance, of the power of a Kola nuclear power plant unit. So there is considerable radioactivity content in a ship reactor, as well.

5.1.2. Erosion and Corrosion

Climatic conditions are tough along the Kola seashore. For instance, buildings and structures deteriorate quickly due to the repeated freezing and melting of water and to moisture penetrating openings and cracks. Salt in the atmosphere and droplets of sea water blown into the air during stormy weather speed up the corrosion rate of metals. Under such conditions, spent fuel in long-term storage, as at Andreev Bay where the fuel has been stored for decades, is at risk in the long run. The buildings and structures housing fuel stored indoors may deteriorate, as may the containers holding the fuel, whether they are stored indoors or outdoors. At the same time, the fuel itself and the cladding can corrode from the inside, all of which may result in leakage of radioactivity to the ground and the sea.

Leakage into the sea is bad enough, but release to the atmosphere can be considered a greater threat. I am therefore more concerned about accidents associated with spent fuel than with long-term leakage to the sea. Accidents such as criticality, fires, and other risks which could bring radioactivity to the atmosphere are of particular concern.

5.2. ASSESSING RISK

So, there are risks associated with the storage of spent nuclear fuel, and the magnitude of the risk must be assessed in some way. It is difficult to assess the risks since very little information about the fuel has been released, although some information is available. Over the years, detailed information has gradually emerged, and there is now sufficient information available to make some evaluations about the magnitude of the hazards.

5.2.1. Sevmorput

As one example, in 1990, the Norwegian Nuclear Energy Safety Authority[1] received a safety report about the nuclear-propelled ship *Sevmorput*—a cargo ship with icebreaking capacities that has a reactor nearly identical to icebreaker reactors—where the core is described and drawings of the reactor core are depicted. There are no drawings of the fuel assemblies, but there is detailed information about the core. Table 2 contains specifications concerning the core of the *Sevmorput* reactor. With the information given in the safety report, it is possible to deduce that the fuel assembly has 55 positions for fuel rods, which makes it possible to describe the fuel fairly well. Some of the positions are occupied by tubes containing gadolinium, which absorbs neutrons and therefore acts as burnable poison in the reactor core.

---

[1] In 1993, the Norwegian Nuclear Energy Safety Authority and the Norwegian State Institute for Radiation Hygiene were merged into one organization—the Norwegian Radiation Protection Authority (NRPA). On 1 January 1993, the NRPA legally became the competent authority for nuclear safety and radiation protection in Norway.

The safety report provides sufficient information to calculate the reactivity in the reactor, and the reactivity calculation provides a basis for assessing the risk of an uncontrolled nuclear chain reaction (criticality accident).

TABLE 2. *Sevmorput* Reactor Core Data.

Fuel Composition	Uranium-zinc alloy
Mass of uranium-235	150.7 kilograms
Height of active part of core	1000 mm
Described diameter of core	1212 mm
Number of fuel assemblies	241
Enrichment level of fuel	90 percent
Diameter of assembly	60 mm
Fuel pin spacing in fuel assembly	7.0 mm
Outer diameter of fuel pin	5.8 mm
Heat release area of core	233 m^2
Passage area for heat transfer agent in core	0.26 m^2

6. International Cooperation

Norway feels that it is a Russian task to describe the situation and to discuss the possibility of accidents, which the Norwegians think is the greatest concern. Norway is more anxious about the risk of accidents than the present level of radioactive contamination in northern areas because the contamination is currently low, while many accidents have already occurred. Six reactors with fuel have been dumped at the base close to the shore of Novaya Zemlya. These reactors have suffered accidents. There are also reports of nuclear submarines with damaged reactors along the Kola coast—perhaps five, six, seven, or even eight. There are ten reactors without fuel reported to be dumped, including reactors that have undergone accidents. Damaged fuel at Andreev Bay and Gremikha is also a sign of reactor accidents. All of this bears witness to the fact that accidents have occurred. The Norwegians are concerned about that and believe that it is a Russian task to describe the sites—all sites—and that there must be full openness about their condition. The Norwegians need a description and written safety analyses from the Russian side. Norway would very much like to help to bring such analyses forward. However, analysis can only be based on information confirmed by Russian authorities. Information which Russian authorities are unable or unwilling to confirm is of little value for this work.

Analysis of the possibilities of criticality, fires, explosions, sabotage, and mechanical failure is a necessity, as is discussion of the contamination hazard associated with the possibility of slow release. The Norwegian side is prepared to bring in some experts and organizations with calculation codes for theoretical calculations in these matters.

Concerning criticality analysis, it has now been suggested that the Norwegian firm Scanpower work together with the Norwegian Radiation Protection Authority[1] to make some calculations about Russian ship reactors with respect to criticality and radioactivity content. The Norwegians would very much like to cooperate with Russian institutes and officials in such calculations and will propose a joint research program to begin studying first criticality safety for icebreaker reactors, and then submarine reactors. The program proposal will include studies of criticality issues for spent fuel handling operations, including transport and storage.

CHAPTER 13
TIME-RISK METHODOLOGIES FOR EXAMINING REMEDIATION TECHNOLOGIES FOR WASTE CONTAMINATION SITES

EDWARD A. MCBEAN, FRANK A. ROVERS, DARRELL E. O'DONNELL
Conestoga-Rovers and Associates Ltd.
Waterloo, Ontario, Canada

1. Abstract

The following is a discussion of elements of the temporal variability of exposure risks and the need to consider the uncertainties of exposure risk estimates. As an aid in characterizing the uncertainty and providing the means to incorporate it into the selection among maintenance/remediation approaches, the utility of time-risk curves is described.

As an example of the application of these methodologies, the techniques for mining and subsequent disposal of process wastes have evolved significantly in recent decades. Nevertheless, the decommissioning of tailings sites continues to raise important concerns about minimizing long-term care and maintenance problems. In response, a number of procedures are being explored as permanent methods to permit "walkaway". Discussions are presented for several of these methods.

2. Introduction

The process of considering risk or—in a more formal sense—risk assessment, seeks to estimate the likelihood of the occurrence of adverse effects in humans and ecological impacts within an ecosystem due to exposures to chemical, physical and/or biological agents. The intense interest currently being demonstrated with respect to risk assessment is driven partly by the potentially sizable costs associated with environmental protection and/or remediation once problems have already developed. Thus, significant interest exists in determining whether portions of these costs can be avoided in situations where the exposure risks are considered acceptable. As a result, the consideration of risk is mandatory in virtually all areas of environmental assessment.

In a typical scenario in which there is an exposure, contaminants may be transported via one or more media—including air, soils/sediments, surface water, and groundwat-

E.J. Kirk (ed.), Decommissioned Submarines in the Russian Northwest, 141–158.
© 1997 *Kluwer Academic Publishers.*

er—to potential receptors through inhalation, dermal contact, and/or ingestion. The exposure assessment aspect of the risk assessment must then characterize the physical and exposure setting, including:

- the identification of significant migration and exposure pathways;
- the identification of potential receptors;
- the development of exposure scenarios, including the determination of current and future exposures; and
- the estimation of chemical intakes for all potential receptors and significant pathways of concern.

In addition, the temporal variability of risk may be very relevant when analysts are attempting to select a preferred alternative among environmental control and/or remediation technologies.

As a consequence of such features as temporal variations in source releases, temporal limitations on the assimilative capacity of aspects of the environmental media, and the temporal variability of the absence or presence of human receptors, exposure risks may vary with time. Therefore, consideration of the temporal variability of the exposure risk is a very relevant dimension in making decisions about many environmental problems.

3. Basis of Risk-Time Curves

The use of cost-time information in the areas of environmental protection and remediation has been standard practice for many years. Typically, cost-time curves themselves have been considered only peripherally since the temporal variability of the costs is modified to a present value and/or translated into an equivalent annual cost using an economic discount rate. The equivalencing concept is used to remove individual temporal variations, which places alternatives onto an equal basis and allows them to be compared.

Risk-time curves are of a similar nature, except that the curves indicate the temporal changing levels in terms of risk as a function of time for each of the control and/or remedial alternatives. The quantification of the risk-time curves necessitates the use of environmental pathways models to characterize the migration of constituents from the source to the point of exposure. These quantifications must reflect the temporal variations associated with the release mechanisms since not all of the various release mechanisms are necessarily functioning at all times.

Aspects of incorporating risk-time curves into decision-making on environmental remediation issues have been demonstrated in [3].

4. Uncertainty in Risk Assessments

The need to consider the uncertainty of risk estimates when applying risk assessment procedures is of paramount importance. Although uncertainty analyses have some-times been viewed as the last step in the risk characterization process, they are a fun-damental element of each component of risk assessment, and the results for each com-ponent must be presented together with an explicit statement of the degree of confi-dence. These measures provide the bases for estimating the degree of confidence in the risk assessment itself, which is crucial.

Consideration of the uncertainty must include numerous dimensions, including:

* the assignment of probability distributions for the contaminant source-loading functions; and
* parameter assignments to be used in migration pathways models.

Modeling approaches that incorporate uncertainty information into the input parame-ters include the Monte Carlo sampling procedure and the Latin Hypercube sampling procedure (see, for example, [2,4]).

The following sections provide an example of how these principles can be applied by describing the characterization of uncertainty and the means of incorporating it into the process of selecting maintenance/remediation approaches in reference to radioac-tive mining and process wastes.

5. Control of Migration Of Mining and Mill Waste Components

Significant improvements in the technology of mining and milling, waste manage-ment, and environmental restoration have been developed in recent decades. However, as reported in [1], there are currently about 200 million tons of uranium mine tailings and mine waste rock on the surface in Canada, which illustrates that sizable quantities of wastes exist that must be dealt with. The geographical distribution of these wastes is indicated in Table 1.

TABLE 1. Uranium Mine Wastes in Canada

	Tons (million)
Northwest Territories	1
Saskatchewan	26
Southern Ontario	6
Northern Ontario	165
Total	198

Reference:[1]

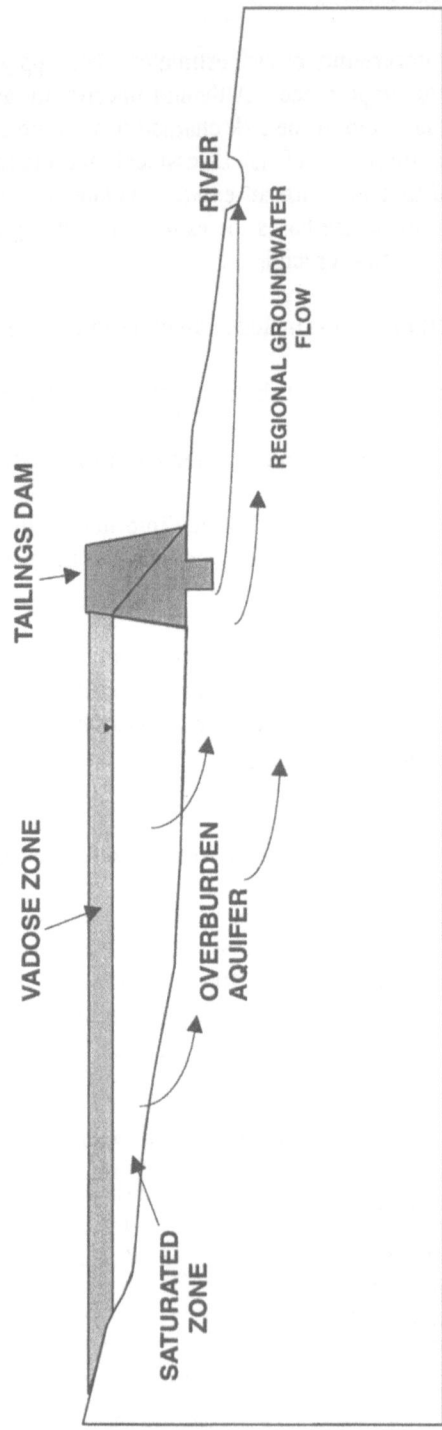

Figure 1. Tailings Impoundment Configuration

145

These tailings are typically stored behind embankments. Figure 1 provides an example of this in a schematic depiction of the configuration typically used for the Elliot Lake, Ontario area. These tailings are underlain by Precambrian sandstones.

The uranium mine wastes at Elliot Lake, for example, include low-grade uranium ore (less than 0.1 percent uranium), existing mainly as brannerite mineralization in a pyrite-rich siliceous deposit. This means that the tailings are strong acid generators from the residual pyrite, which results in high sulfide content in the leachates and low pH. Because of the acidic character of the leachate and the exposure pathway by which the radioactive materials may induce unacceptable exposure to humans and the environment, there is interest in improving disposal technologies.

The disposal of waste rock and mine tailings from sulfide bearing rock creates one of the major environmental problems associated with mining. Sulfide oxidation through various mechanisms creates acid mine drainage which has a significant impact on the environment. Acid mine drainage can lower pH and increase heavy metal concentrations in aquifers and surface water systems. Furthermore, since tailings are likely the source of regional aquifer recharge, the impact on the environment will continue for many years unless it is remediated.

The problems associated with mine tailings are exacerbated when waste rock is placed on top of the tailings, as was a frequent historical practice. The high volume of air in the waste rock piles allows for ready oxidation of sulfides, which generates heat. Convection draws in more air, replenishing the oxygen supply, which allows for further oxidation of the pyritic minerals in the waste rock. The higher temperatures in the waste rock increase the rate of pyrite oxidation, creating a rapid buildup of oxidation products and high temperatures (temperatures exceeding 50° Celsius are not uncommon in waste rock piles). The oxidation products remain in the waste rock until they are flushed from the pile by rainfall, creating extremely acidic conditions over a very short time period.

Consequently, the objectives of focusing on improved disposal to tailings ponds include:

- minimization of acid generation;
- minimization of radiation exposures; and,
- maximization of long-term stability of impoundments.

In Canada, as elsewhere in the world, governments are neither willing nor capable of implementing costly remedial actions. Consequently, a number of alternatives are being considered with a view towards avoiding long-term costs and environmental damage. The alternatives include:

5.1. WATER COVER

This procedure utilizes runoff from adjacent lands and internal dikes to retain a water cover. The use of water cover is an important technique for preventing intrusion, radon evolution, and acid generation by minimizing oxidation opportunities. It will be examined more fully below.

5.2. LIME TREATMENT

In this procedure, lime is added to neutralize the acidity of the leachate. The process of milling uranium bearing ore involves the use of acid to extract the uranium. This use of acid consumes any natural alkalinity in the ore. Thus, there is little, if any, alkalinity in the tailings. The addition of lime ($CaCO_3$) to the tailings reestablishes their alkalinity. The buffering capacity of the tailings pile is then able to neutralize the acid generation capacity of the pyritic tailings, provided that enough lime is added. Generally, a three to one ratio of acid neutralization capacity to acid generation capacity is employed.

5.3. GLACIAL TILL CAP

Closure of waste rock piles and tailings impoundments often involves the installation of caps with low permeable soil layers to minimize the amount of infiltration. However, the placement of caps is expensive, and caps must be monitored to ensure that their integrity is not compromised.

5.4. DISCHARGE TREATMENT IN MAN-MADE WETLANDS

This procedure involves drainage to man-made wetlands. Recent research by the Canadian government partnered with various mining companies indicates that biological treatment of acid mine drainage runoff in man-made wetlands is a viable alternative. Essentially, this method involves creating a wetland environment into which the acid mine drainage runoff is allowed to leach at a controlled rate. The most promising system uses a floating canopy of bullrushes that creates a reducing zone below the water surface where microbial reactions reduce the sulfides, thus neutralizing the acidity. This system, when properly designed and implemented, results in a cost-effective long-term solution that requires minimal maintenance.

6. Tailings Ponds

Mine tailings impoundments are usually large ponds into which a slurry of tailings are pumped. While the tailings are underwater, oxidation is limited due to the prevalent reducing conditions. However, when an impoundment is closed, the water level usually decreases, exposing the top layer of tailings to direct contact with oxygen, which begins the oxidation process.

Many sites in Canada have reported very acidic conditions in runoff from waste rock piles and tailings impoundments. Measurements of pH as low as 1.9 have been recorded at some sites. Acidic conditions such as these destroy the habitat of flora and fauna that come into contact with the runoff.

Older mine sites (abandoned or still operational) typically have either existing acid mine drainage problems or the potential to generate problems in the future. Aquifers that already have acid mine drainage contamination can either be left as is, or migration can be minimized by very costly pump and treat systems that create hydraulic barriers to stop the travel of contaminated acid mine drainage water. Bactericides can be injected into aquifers to reduce the bacterial catalysts, but this treatment does not last very long, and it becomes prohibitively expensive if used as a long-term remediation method.

Newer mine sites can benefit from preventative measures designed to avoid the problem of acid mine drainage. Currently, limed tailings impoundments are often utilized.

7. Importance of the Redox Reaction to Minimize Migration

Redox reactions involve organic matter and the solutes in water. In groundwater, in effective contact with sediments, redox reactions involve the solid phases. Decreasing the redox potential can cause the precipitation of uranium (i.e. uranium moves into the groundwater flow system). When uranium reaches reducing conditions, it precipitates.

Redox reactions in pyritic mine tailings can create acid mine drainage (AMD), which is a significant problem due to its direct and indirect impacts on the environment. The major direct impact is the decreased pH of water that is discharged from the tailings impoundments into ambient surface waters and groundwaters. The indirect impact of concern to uranium mine tailings is the mobilization of heavy metals (including uranium, thorium, lead, and radium) that result from low pH levels.

Under oxidizing conditions, pyrite is oxidized to create acidic conditions. Several steps are involved in the oxidation of pyrite, but the net reaction proceeds as follows:

$$4FeS_2 + 15O_2 + 14H_2O \rightarrow 4Fe(OH)_3 + 8SO_4 + 16H^+$$

This equation indicates that under ideal conditions, one mole of pyrite can generate four moles of H^+. In typical uranium mill tailings, the alkalinity has been removed in the milling process, which means there is minimal capacity for buffering the acid generated by the oxidation of pyrite.

The rationale, then, behind procedures such as maintaining a water cover is to minimize the opportunities for oxidation processes. This, in turn, minimizes the acid drainage aspects and the migration opportunities for the radioactive materials. The controlled drainage approach, as schematically depicted in Figure 1, is designed to allow tailings to drain and consolidate. The drainage enhances the structural stability of the tailings, making them less prone to structural failure. However, if drainage is allowed to occur, atmospheric gases enter into the vadose zone, allowing oxidation. Precipitation water then carries the vadose zone reaction products, allowing their migration. Thus, it is important to seek consolidation through seepage while maintaining a water cover. The intent of maintaining an impoundment is that the lake bottoms will create a natural zone in which oxidation is unlikely to occur while leachate percolating from the impoundment will maintain reducing conditions.

A relevant feature of migration characteristics is that of roll-front deposits. Specifically, when oxygenated water starts to flow through an aquifer in which conditions were initially reducing, a redox front may develop between oxidizing and reducing environments. The front will move in the direction of groundwater flow but at a rate that is much slower than that of the water. This prevents rapid intrusion of the precipitation, which can initiate oxidizing conditions. Since uranium is insoluble under reducing conditions and soluble under oxidizing conditions, as the oxidation front advances, any uranium elements present in the aquifer are dissolved. The moving groundwater migrates through the front into a reducing environment, where the elements immediately reprecipitate.

Given all of the above, interest is focused on maintaining reducing conditions in the tailings pond and minimizing oxygen intrusion.

8. Discussion of the Model and Estimated Radiation Doses From Tailings Impoundments

The tailings impoundment model used is a simple one. Simplicity of the model permits easier uncertainty analysis and modification of the model to represent the various site actions. A simple diagram of the tailings pile breakdown is depicted in Figure 2.

The tailings pile itself has been broken down into three vertical sections. The first section is the vadose zone, which was assigned a uniformly distributed thickness between 0.5 and two meters. The underlying saturated zone is approximately eight meters thick. It was used to represent the mixing zone. Below the mixing zone is a thin (less than one meter) saturated layer, which discharges into the underlying aquifer.

The aquifer under the tailings pond is thin and sandy with a hydraulic gradient ranging from 0.04 meters per meter to 0.06 meters per meter. This aquifer discharges into a short (2.3 kilometer-long) river.

The surface water system consists of the river and a small lake. The catchment areas and other relevant information about the surface water system are shown in Figure 3, which also indicates the location of the two receptors.

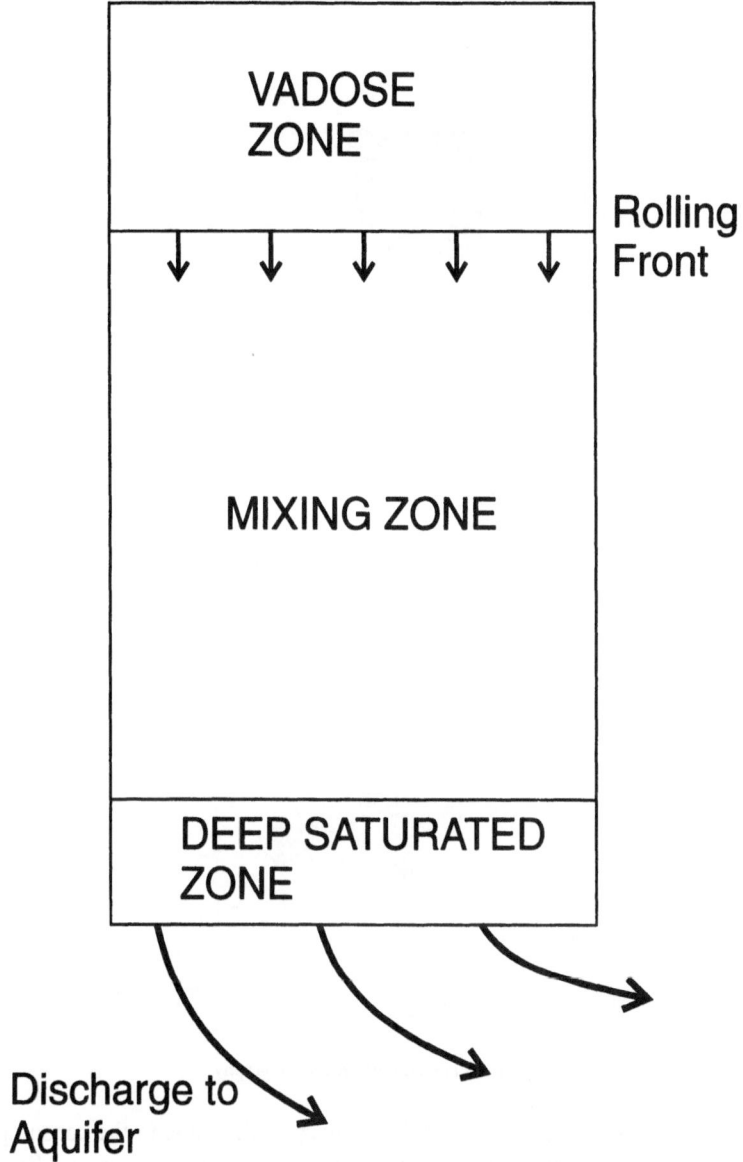

Figure 2. Cross Section of Tailings Impoundment

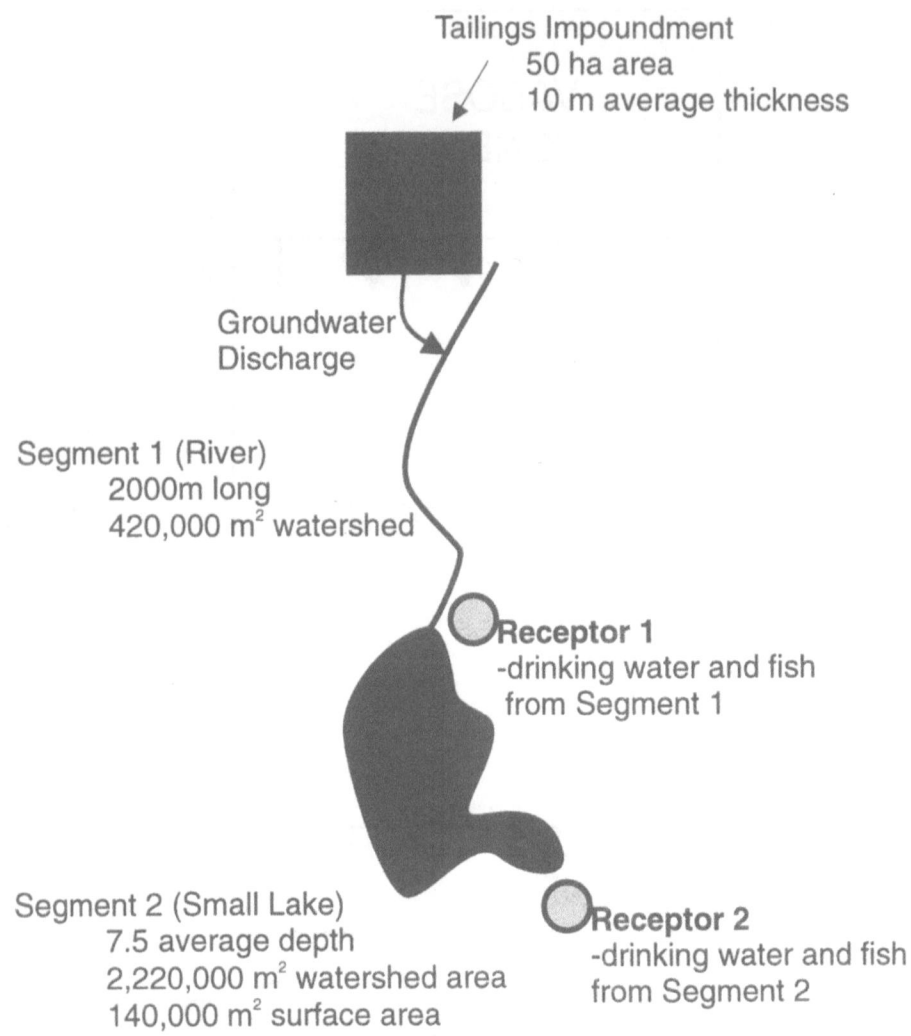

Figure 3. Plan View of Model Geography

The two receptors were selected to model the effect on dose due to the consumption of drinking water and fish from the various contaminated surface waters. Receptor 1 consumes drinking water from the river segment whereas Receptor 2 consumes drinking water and fish from the lake. Doses were estimated for the do-nothing site alternative and four other site alternatives: waste rock cover, lime treatment, glacial till cover, and water cover. Results of the model are described below.

Figure 4 shows the general time-dose trend, indicating that a high dose of radiation occurs in the first 100 years from the closure of tailings impoundments. This dose decreases for the next 200 years, at which point the dose begins to increase again, peaking at 700 years. After 700 years have passed, the dose once more begins to decrease.

The initial dose is due to thorium isotopes and lead-210. The acidic conditions created by the oxidation of pyrite dissolves heavy metals, making them more mobile. However, the aquifer media typically retard the transport of uranium and radium, which allows for the relatively rapid migration of thorium and lead, while maintaining a slow leach of uranium and radium. The dose from thorium is due to the lower retardation of thorium in the aquifer, which means that thorium migrates far more quickly than uranium and radium. Lead is transported at a faster rate than uranium and radium but a slightly slower rate than thorium. However, the rapid and energetic decay of lead-210 (96 percent of it decays in 100 years) more than compensates for the longer travel time. Thus, lead-210 contributes the largest dose initially, due to its relatively rapid transport and its rapid radioactive decay. The rapid radioactive decay also accounts for the quick drop in dose due to lead-210 over the first 100 years from closure.

After the lead-210 is consumed by radioactive decay, thorium is the major isotope contributing to the dose. For about 300 years, the dose continues to decline as thorium decays until enough radium-226 has been created that the dose from radium-226 surpasses the dose from thorium. The dose from radium-226 increases until the amount of radium being produced in the tailings decreases to a level where the infiltration and migration of the radionuclides surpasses the radium-226 generation rate. At this point, the overall dose begins to decrease.

The contamination pathways that the radionuclides follow vary with time, as shown in Figure 5. The only pathways of concern in this circumstance are the consumption of drinking water and fish from the impacted water segment. Other pathways—such as inhalation, radon, vegetable, and groundshine—contribute a negligible portion of the total dose received at either of the test locations. Drinking water is the major pathway during the first 400 years, at which point the dose due to fish consumption becomes important. The drinking water pathway is still the largest dose pathway, but the fish consumption pathway rises from 100 years until 700 years, indicating that bioaccumulation of radionuclides impacts the dose pathway.

Four alternative site actions were analyzed for dose. The alternatives (waste rock cover, glacial till cover, lime treatment, and water cover) were compared to the do-nothing alternative to compare time-dose trends. The dose expected at Receptor 1 (drinking water and fish sourced from the river) is depicted in Figure 6. The Receptor 2 (drinking water and fish sourced from the small lake) dose is depicted in Figure 7. Since Figures 6 and 7 display similar time-dose trends, except in the magnitude of the dose (due to dilution), they will be discussed together.

The general trend in the dose curves is similar to the trend explained above for the component pathways. The dose tends to start at a very high level, dropping in the first 100 to 150 years. The dose levels out until 300 years have passed, at which point it begins to rise, until it peaks at 700 years.

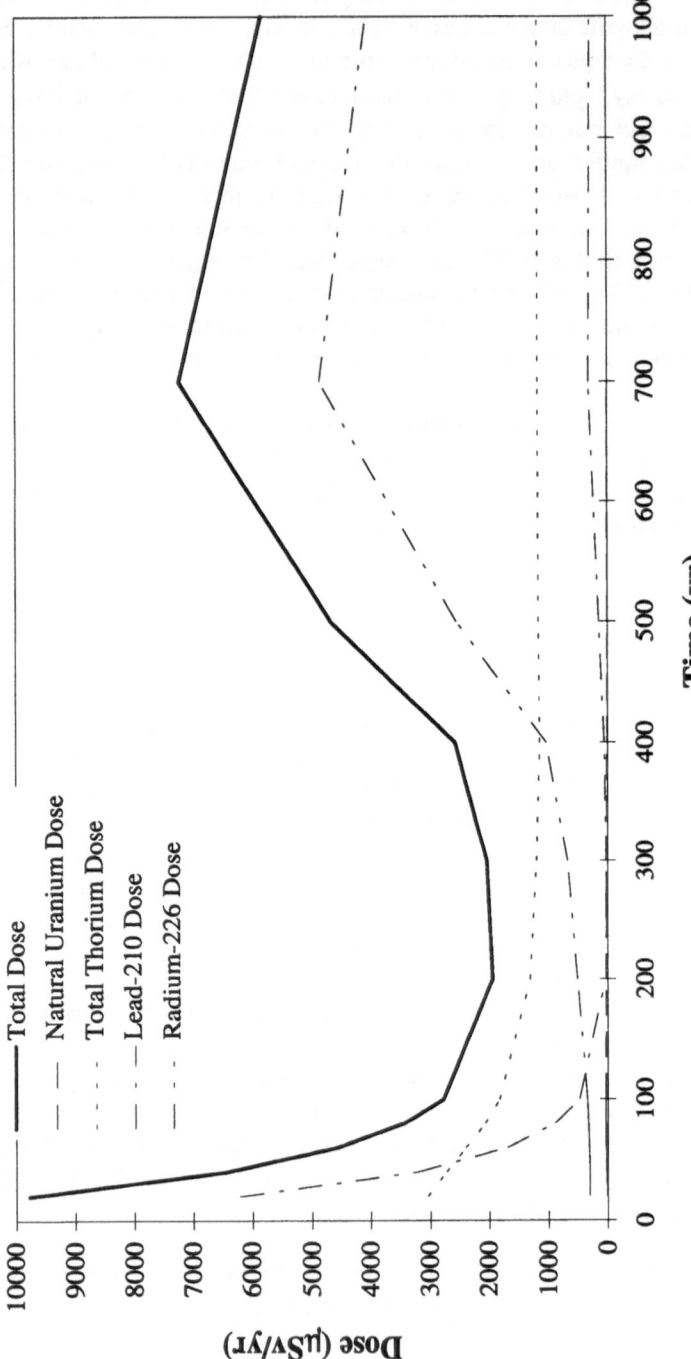

Figure 4. As Is Dosages—Radionuclide Components at Receptor 1

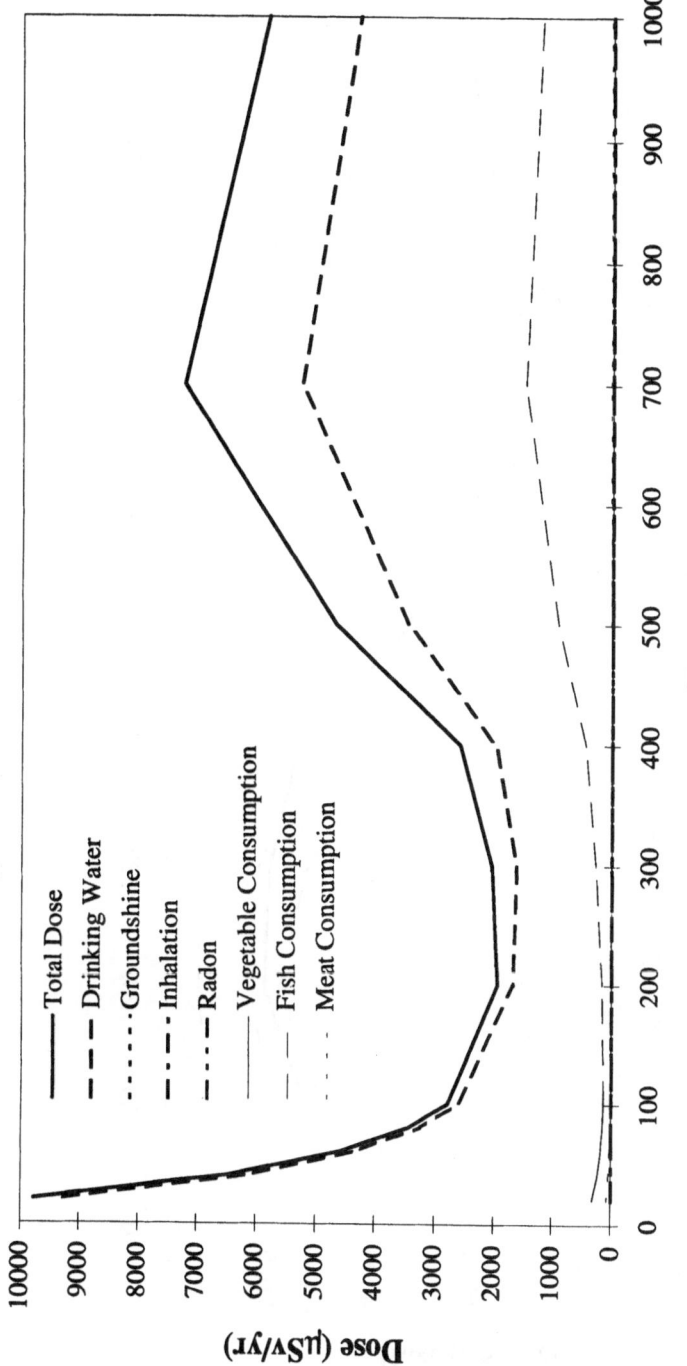

Figure 5. As Is Dosages—Radionuclide Pathways at Receptor 1

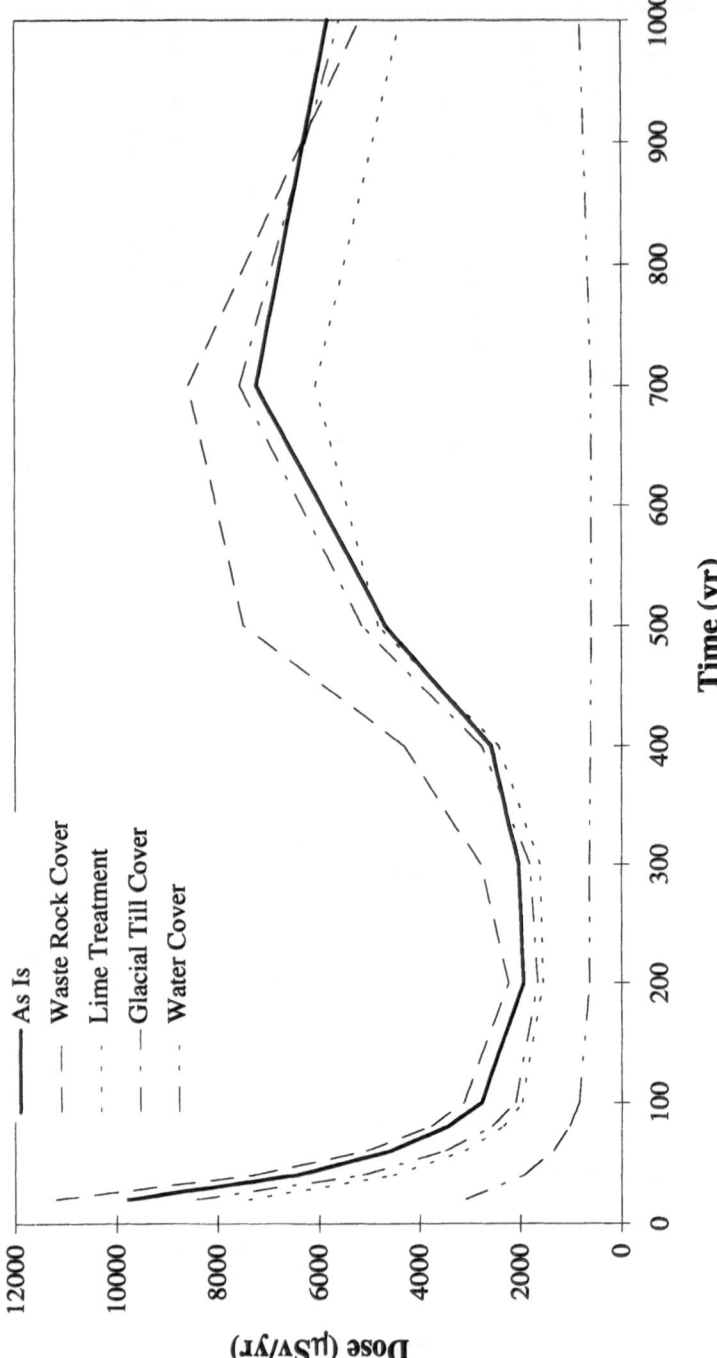

Figure 6. Site Action Dose Graph—Receptor 1

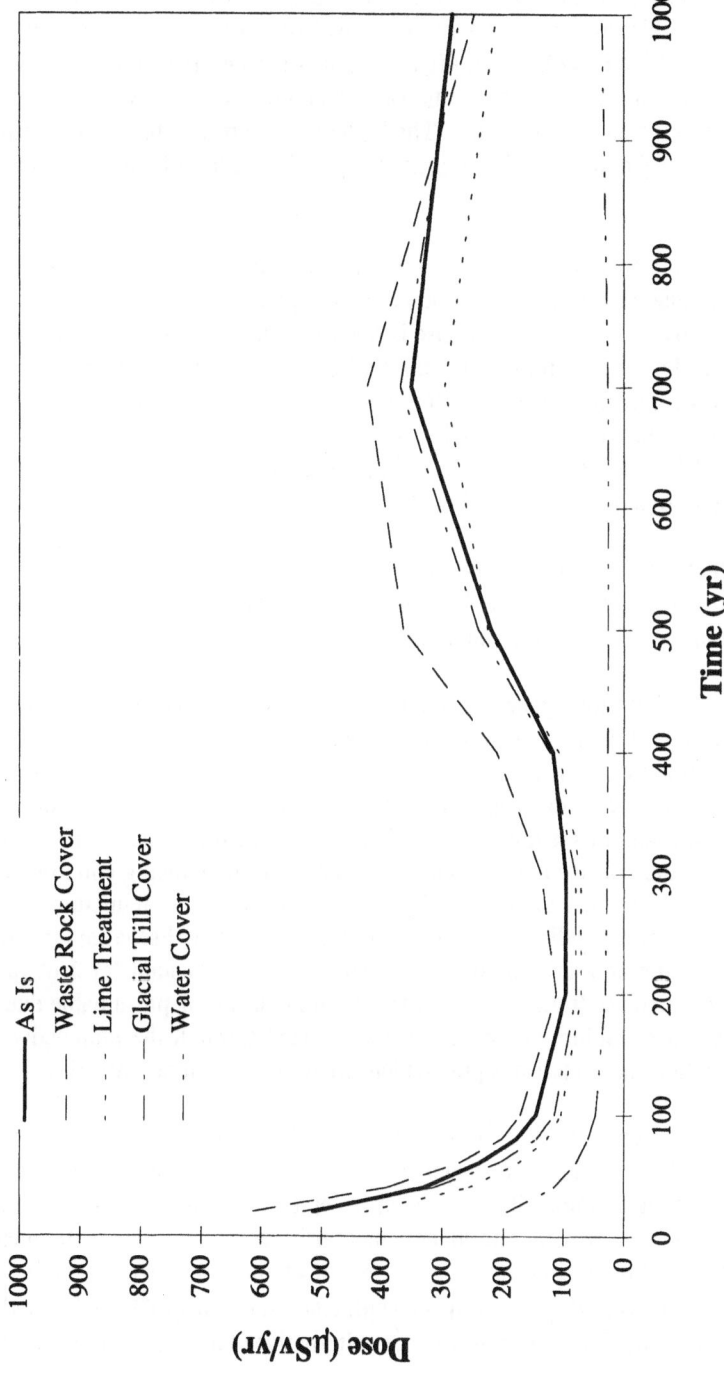

Figure 7. Site Action Dose Graph—Receptor 2

The highest dose is generated by the waste rock cover alternative. The high dose was anticipated, and this alternative was never considered to be a remedial option. The waste rock alternative is included in Figures 6 and 7 because the storage of waste rock on top of closed tailings impoundments was a common historic practice, so the dose due to this storage method is relevant. The high dose can be attributed to the complete capture of rainfall (i.e. no runoff) and lack of vegetation cover on top of the waste rock pile.

The do-nothing alternative causes the second-highest dose in the first 300 years. After 300 years, the glacial till cover alternative has a higher associated dose than the do-nothing alternative. This change in dose is due to the larger remaining amount of radionuclides in the tailings impoundments that exists if a glacial till cover is placed over the tailings impoundment. The low permeability cap generates less infiltration into the tailings impoundment and hence a lower migration rate into the aquifer. Thus, the available radionuclides in the tailings impoundment when a glacial till cover is used is higher than in the do-nothing alternative.

The lime treatment alternative causes the second-lowest dose of the alternatives modeled. This low dose is due to the reduction in the amount of acid generated in the tailings impoundment, which lowers the migration rate of the radionuclides.

The water cover alternative generates the lowest and most consistent dose. The initial dose begins at less than 40 percent of the do-nothing alternative. The dose drops off rapidly and maintains a low level. It should be noted that the other four alternatives modeled follow an increasing trend from 300 to 700 years from closure, while the water cover option maintains a low level, without an increasing trend from 300 to 700 years. The low and consistent dose can be attributed to the reducing conditions at the surface of the tailings that are created by the water cover. The reducing conditions limit the mobility of the radionuclides in the tailings pile. The discharge from the impoundment into the aquifer contains a low and relatively constant level of radionuclides. A small increase from 700 to 1000 years from closure is presumable due to the increased amount of radium and other isotopes created through the radioactive decay of the radionuclides in the tailings pile, which would then leach out over time.

To demonstrate the circumstance of uncertainty in tailings pond concerns and radioactivity, consider the configuration depicted in Figure 8, which shows the dose versus anticipated site action complexity and cost at Receptor 1. Using 100 trials in a Monte Carlo simulation, a 95 percent confidence interval is shown along with the mean anticipated value. Although the mean value for all of the alternatives remains below 1000 μSv per year, the 95 percent upper confidence interval (95% UCL) of the do-nothing, lime treatment, and glacial till cover alternatives far exceed the 1000 μSv per year threshold.

The water cover option is clearly the most acceptable site action as far as the long-term radioactive dose is concerned. However, the cost and complexity of implementing this

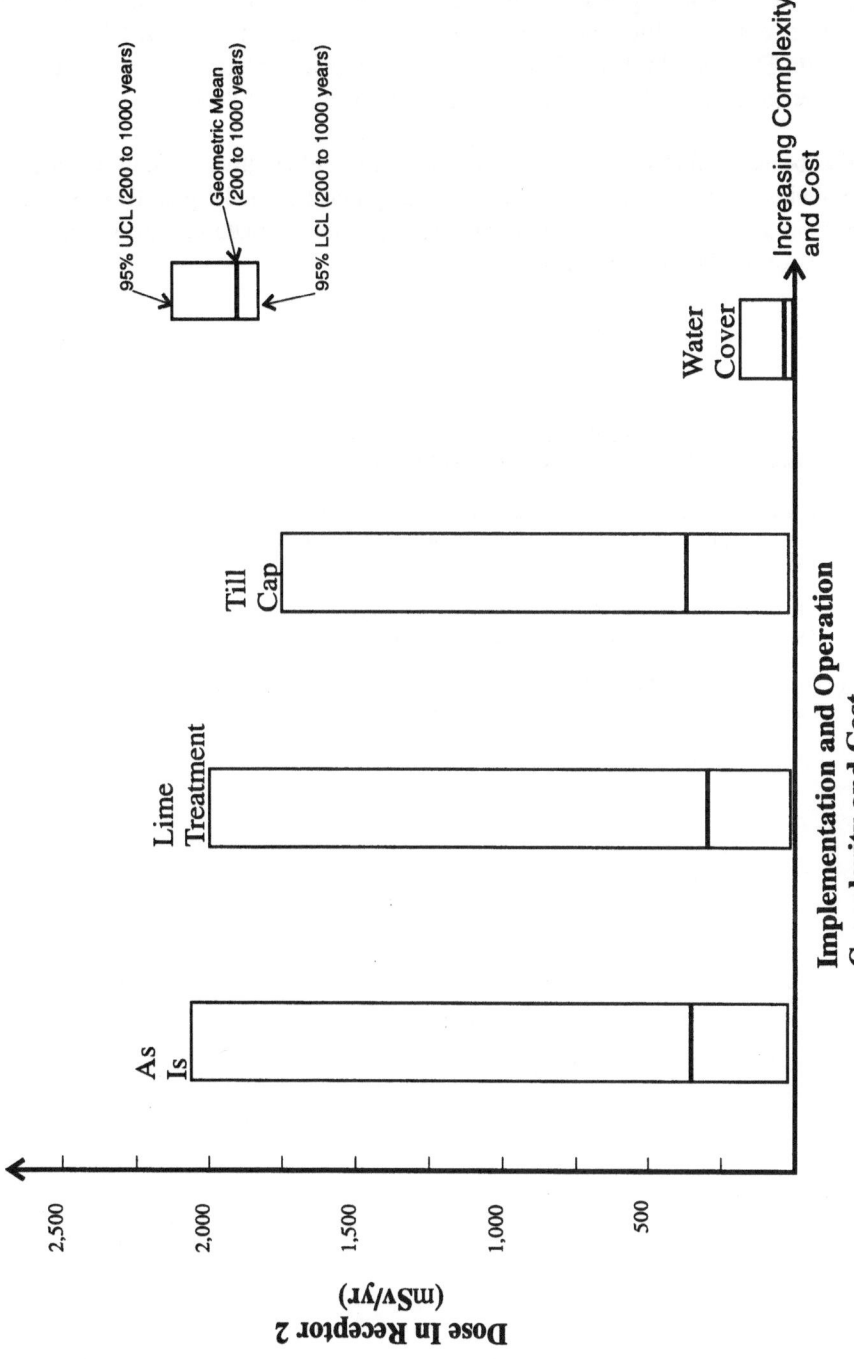

Figure 8. Risk-Cost Depiction of Four Alternatives

option must be weighed against the risk reduction that the option provides. If the cost is excessive, either the lime treatment or the do-nothing alternative would have to be considered. The glacial till cap alternative would not be considered because it causes a higher average dose yet it is more complex and expensive to implement.

It should be noted that the doses and the costs of the alternative are only estimates. Each individual site should be assessed separately and the model modified to fit the circumstances. When more in-depth data are available for making a decision, more reliability can be placed upon the results.

References

1. Feasby, D.G. (April 1996) *Affordable Technologies for Remediation of Sites Contaminated by Uranium Mining and Milling Activities*, presented at the I.A.E.A. Workshop in Ukraine.

2. Iman, R.L. and Conover, W. (1980) Small Sample Sensitivity Analysis Techniques for Computer Models, with an Application to Risk Assessment, *Commun. Stat. Theor. Methods*, **Series A 9, no. 7**, 1749-1842.

3. McBean, E. and Rovers, F. (1995) Utility of Risk-Time Curves in Selecting Remediation Alternatives, *Journal of Waste Management and Research* **13**, 167–174.

4. Murray, G., McBean, E. and Rovers, F. (Spring 1996) Risk-Based Engineering Design for a Landfill Leachate Collection and Liner System, *Ground Water Monitoring and Remediation*, 139-146.

V. SUMMARY AND CONCLUSIONS

CHAPTER 14
INTEGRATION OF TECHNICAL PROJECTS WITH LARGER POLITICAL AND ECONOMIC CONTEXTS

ELIZABETH J. KIRK
American Association for the Advancement of Science
Washington, D.C., United States of America

1. Introduction

The first chapter of this volume, written by Drs. Lystsov and Khlopkin, identified the major problems facing Russia today regarding the safe disposition of radioactive wastes produced by the operations and decommissioning of Russian nuclear submarines in the Arctic North. Many of the remaining chapters then discuss existing procedures and technologies that can be applied to the removal, processing, storage, and transport of spent nuclear fuel and liquid and solid radioactive wastes. These analyses come after many studies identifying the sites and extent of contamination in the region. [5,6,7,9] One of the conclusions reached at the NATO Advanced Research Workshop in Kirkenes, from which these papers are taken, was that many of the technologies already exist to resolve current radioactive waste problems in the Arctic North.

It would seem an easy task, therefore, to simply apply these technologies to the problems at hand. The job is made more challenging, however, by the many political and economic constraints that must be overcome for the successful execution of all the activities involved in the disposition of wastes produced by the decommissioning of nuclear submarines.

This final chapter attempts to put these technical issues in larger political and economic contexts and examines the progress that has been achieved to date in prioritizing waste handling, reduction, transport, and storage requirements and in providing the cooperative teams and financial resources needed to meet these requirements.

2. Russian Perspectives

The Russian ministries and other institutions tasked with the safe and secure handling of radioactive wastes resulting from nuclear submarines are coming to the negotiating tables on these issues from at least three perspectives:

E.J. Kirk (ed.), Decommissioned Submarines in the Russian Northwest, 161–174.
© 1997 *Kluwer Academic Publishers.*

- national and international environmental security concerns;
- international security concerns; and
- national security concerns.

The first concern focuses on the assurance that the wastes produced by the decommissioning of nuclear submarines will be handled and disposed of properly and will pose no threat to human or other populations or the environment as a whole. Through the London Convention, which prevents ocean dumping of nuclear wastes, and through many bilateral and multilateral environmental agreements, Russia has assumed responsibility for preventing accidental or purposeful releases of radioactive wastes into the environment. Although Russia has not formally agreed to all aspects of the London Convention (i.e. the dumping of liquid wastes), it has agreed, in principle, to avoid such dumping assuming it can get international support to expand existing waste treatment and storage facilities. Other environmental security activities, in general, also focus on cleaning up wastes on existing military and civilian sites.

The second perspective comes from the military requirement to decommission submarines, either because of attrition or because of international security agreements such as START II. The removal, dismantlement, and destruction of nuclear weapons on board submarines and the decommissioning of submarines themselves pose enormous problems to the Russian Navy, both in the Arctic North and the Russian Far East, due to the lack of adequate facilities to handle all of the associated wastes. These waste handling problems are additionally complicated by threats of theft of nuclear materials (especially plutonium and highly enriched uranium) and the exportation of know-how associated with nuclear submarines and weapons. Given severe budgetary constraints, the Russian Navy must decide whether to spend its limited resources on operational maintenance of the Navy and submarine construction or on handling the wastes that will result from dismantling submarines that have already been taken out of service in order to meet the terms of international agreements. Thus, international security concerns brought about by bilateral or multilateral agreements can come into direct conflict with national security concerns for maintaining an adequate defense. They may also coincide in that the same dismantling and handling issues arise from START II as from the decommissioning of submarines that are past their realistic operational life.

Finally, there is an additional national security concern brought about by the involvement of international participants in some of these projects; namely, there is mistrust of the intentions of some partners and suspicion that they may be cooperating for industrial espionage or other intelligence-collecting purposes. The Russian Nuclear Navy, as other nuclear navies, has been reticent to open its bases to foreign scrutiny.

It is understandable, therefore, that the Russians have approached cooperative projects on the remediation and disposition of wastes produced by the decommissioning of Russian nuclear submarines very cautiously.

2.1. RUSSIAN ACTORS

Many major and minor ministries and institutions are involved in the several processes concerning the removal, transport, reduction, and storage of radioactive wastes from submarines. First, the Russian Navy is charged with safely and securely handling spent nuclear fuel and radioactive wastes as they are removed from the submarines. The Navy's Department of Radiation Safety has control over the operations and maintenance of nuclear submarine reactors and the wastes they produce. Until recently, the Ministry of Defense and the Russian Navy have been reluctant to participate in international discussions of Northern Fleet submarine maintenance and decommissioning. However, they have slowly been brought into the process because of the growing international and Russian public awareness of submarine fleet radwaste problems. These issues were first raised in a series of joint U.S.-Russian scientific meetings which took place in the early 1990s. Such practices as ocean dumping were confirmed by the publication of a government report (referred to as the "Yablokov Report" or "White Book"). [4] Finally, nongovernmental organizations such as Greenpeace and The Bellona Foundation have raised the issues of ocean dumping and accidents in the Russian fleet through public meetings and publications.

The second major actor in this process is the Russian Ministry of Atomic Energy (MINATOM), which is involved in the handling of radioactive wastes from both military and civilian activities. MINATOM is involved in all areas of nuclear safety, including the storage of fuel and associated wastes; the safe operation of nuclear power plants; the transport, reprocessing, and storage of all types of nuclear materials (i.e. weapons grade fissile material, nuclear reactor material, and other radioactive wastes); and managing reprocessing and storage facilities like the Mayak Production Association or proposed repositories at Novaya Zemlya.

It is unclear at this point where the responsibilities of the Navy and Ministry of Defense leave off and where those of MINATOM begin. Because of the severe economic constraints placed on both ministries, neither of them has been able to clearly state publicly their coordinated strategic plans for handling spent nuclear fuel and associated radioactive wastes.

The link between the two ministries seems to be a third—the Ministry of Defense Industries (MINOBORONPROM)—which is tasked with building both military and civilian facilities encompassing everything from shipbuilding and maintenance facilities, to storage casks for wastes, to reprocessing and storage facilities. There are several state industrial complexes involved in the removal and transport of spent fuel, the treatment of spent fuel and radwastes, the construction of waste repositories, and the disposal of submarines, to name just a few.

Still other institutions are primarily concerned with issues of environmental monitoring and assessment. The State Committee on Environmental Protection is the primary agency enforcing environmental laws. (However, this committee was recently downgraded from the Ministry of Environmental Protection.) The committee is unclear, at this point, on its stance regarding Russia's dedication to environmental protection. Various institutes affiliated with several government agencies have also been involved in monitoring the environmental effects of ship-based nuclear reactors and in examining the risks involved in ocean-dumped Russian radwastes.

In addition, local authorities like the Murmansk and Arkhangelsk regional governments have also played a role in locating and evaluating contaminated sites. And finally, semi-private or quasi-governmental companies are emerging and they, too, are becoming involved in studies related to waste treatment and ship disposal.

However, it is clear that the three primary actors in Russia dealing with radioactive wastes are the Russian Navy (and the Ministry of Defense), MINATOM, and MINO-BORONPROM, and that as such, they will be the major Russian actors in projects involving the removal, transport, treatment, and storage of wastes associated with decommissioned nuclear submarines.

Two important constraints impede the progress of these institutions toward completing the projects necessary to ameliorate contamination risks. The first—and by far the most important—problem is the lack of financial resources in Russia to build or retrofit existing facilities and equipment to meet present and future demands. The fact that current waste reprocessing and storage facilities are full, that wet storage facilities need to be replaced, and that damaged reactor cores lie in unsafe storage facilities, attests to the fact that inadequate funds have been allocated to do much to alleviate these problems. The national interests of other Arctic countries in solving these problems because of the threats to their own environment have prompted many of them to offer technical and financial assistance, but it will not be enough. Russia will be primarily responsible for the costs incurred in cleaning up its military and civilian sites in the region.

The second problem for Russia is the lack of established interagency linkages for dealing with environmental issues with military involvement. Bureaucrats in the United States constantly complain about the number of interagency task forces and meetings (on such topics as Arctic environmental issues, for example, which involve over 20 different government agencies). However, such meetings establish the foundation for dealing with problems that cut across institutional guidelines and responsibilities. Historically, these task forces or similar arrangements are weak to nonexistent in Russia [10], and this has been especially problematic in Russia's early discussions with other countries on cooperative projects, particularly due to the lack of assurances given to outside partners that they were dealing with principal actors who were actually authorized and had the funds to carry out decisions reached during these meetings.

Over the past two years, there has been an improvement in the cooperation between agencies, but there still seems to be lack of communication and sometimes competition between military organizations and civilian environmental organizations.

3. International Cooperation

International monitoring and assessment of the Arctic environment has been going on for many years and has involved many bilateral and multilateral agreements and activities.[1] By and large, many of the current international monitoring and assessment activities have been carried out, especially by the Arctic Monitoring and Assessment Program (AMAP) of the Arctic Environmental Protection Strategy (AEPS). The AEPS was signed in 1991 by the eight Arctic nations: Canada, Denmark, Finland, Iceland, Norway, the Russian Federation, Sweden, and the United States. Many of these projects focused on general issues of environmental contamination and monitoring.

Concomitantly, nongovernmental organizations (NGOs) like Greenpeace and The Bellona Foundation began to report on possible ocean dumping and other contamination caused by accidents in the Russian Navy. Their reports were confirmed by the Russian Yablokov Commission report in 1992. [4] These activities began to focus attention on nuclear submarines and the potential risks involved with their maintenance and decommissioning—both in the Russian Northwest and the Far East. Since that time, many activities have focused on the remediation of problems and the construction of facilities and equipment for the storage, processing, reduction, and transport of spent nuclear fuel and associated liquid and solid wastes.

International interests in the Russian Northwest are motivated by a desire to protect the fragile Arctic environment and to address the pollution threats to humans, fish, and other organisms. This includes threats not only from nuclear submarines, but also from nuclear power plants and from the harmful heavy metals, chemicals, and other pollutants emitted by both civilian industries and military facilities. The environmental security of the region is threatened by many sources of contamination. Cross-border pollution and pollution of shared river, bay, and ocean waters threaten the health and welfare of the entire region.

A second motivation for international interest in the region is the Russian Federation's requirement to uphold arms control agreements that limit nuclear weapons and delivery systems, call for the destruction of chemical weapons, and prevent the proliferation of nuclear weapons and theft of nuclear materials. These international agreements directly affect the security of Russia's neighbors. Consequently, they are very interested in seeing to it that Russia adheres to these treaties as well as the London Convention on ocean dumping, which was mentioned earlier. These countries are interested in enhancing Russia's capabilities to meet treaty obligations, while at the same time, not assisting in maintaining Russia's war-fighting capabilities.

International interest has shifted recently from identifying and characterizing risks to identifying projects to reduce those risks. Between 1995 and 1996, at least three major

---

[1] For a detailed historical account of these activities, see [7].

studies were conducted which attempted to outline specific projects to safely dispose of radioactive materials from nuclear submarines.

4. Identification of Specific Projects

The first study, published in December 1995, was an international one sponsored by the Barents Council and supported by the Nordic Environment Finance Corporation (NEFCO) which focused on "Proposals for Environmentally Sound Investment Projects in the Russian Part of the Barents Region." [1] This study was prepared by the AMAP Expert Group and covered all types of radioactive and non-radioactive contamination. From this study, five sets of projects were recommended that dealt with radioactive contamination. They are summarized below:

1. Handling and Transport of Radioactive Waste and Spent Nuclear Fuel

*   transporting vessel for spent nuclear fuel;
*   transport ship for transport to Novaya Zemlya;
*   emptying and removal of full waste storage;
*   treatment of liquid radioactive waste with stationary and mobile equipment; and
*   facility for reduction of solid radioactive waste before transport and storage.

2. Regional Storage for Radioactive Waste and Spent Nuclear Fuel (Especially if Not Suited for Reprocessing)

*   storage site at Matochkin Shar; and
*   storage site at Novaya Zemlya.

3. Development of Alternative Techniques for the Decommissioning of Nuclear Submarines

4. Nuclear Safety at the Kola Nuclear Power Plant

*   safety culture, pre-project.

5. Risk and Impact Assessment Including Monitoring Systems

*   risk and impact assessment for man and the environment from military and civilian sources;
*   monitoring system for environmental releases of radioactivity from civilian and military sources;
*   emergency system in the Arkhangelsk region;
*   monitoring system in the Arkhangelsk region; and
*   regional laboratory.

Simultaneously, the Russians were producing a report focusing specifically on the disposal of nuclear-powered submarines. The report also identified specific requirements for the safe decommissioning of submarines and handling of radioactive wastes from them. The Russian Navy, MINATOM, and the Ministry of Defense Industries participated in the study, which was completed in December 1995.

This study was then reviewed by a Norwegian company—Kværner Maritime—and a Russian counterpart—RKK Energia, who completed the report in January 1996. In a meeting in Oslo in February 1996, Norwegian and Russian authorities signed a protocol for future cooperation that will lead to the disposal of 125 Northern Fleet submarines by 2010. [8] They agreed to negotiate and sign a framework agreement for further cooperation and to continue feasibility studies for seven projects. The Norwegian-Russian framework agreement projects are summarized below:

1. Design, construction, and commissioning of a container vessel for spent nuclear fuel.

2. Construction and commissioning of special railway wagons (TK-VG-18) for transporting spent nuclear fuel.

3. Design, construction, and commissioning of a temporary storage facility for liquid radioactive wastes at the Zvyozdochka yard in Severodvinsk.

4. Establishing a mobile facility for concentrating liquid radioactive waste.

5. Design, construction, and commissioning of a temporary storage facility for solid radioactive waste.

6. Measures to enable emptying and further non-use of an environmentally unsafe temporary storage facility for spent nuclear fuel at Andreev Bay, Kola peninsula.

7. Possible assistance in the completion of an intermediate storage facility for spent nuclear fuel from submarines at the Mayak plant in the Urals.

In addition to these broad-based studies, several other international activities were carried out between 1991 and 1996 covering specific projects.

Supported under the Gore-Chernomyrdin Commission (GCC), the U.S.-Russian pollution prevention agreement, and the Norwegian-Russian environmental protection agreement, United States, Russian, and Norwegian representatives began discussions during the London Convention Meeting in 1993 to expand a low-level liquid radioactive waste (LLRW) treatment facility under the control of RTP "Atomflot" in Murmansk. This facility treats primarily wastes produced by the nuclear icebreaker fleet, but also low-level liquid radioactive wastes from naval reactors. Presently, it cannot accommodate the waste backlog or future waste to be generated by the decommissioned submarines and the operational fleet. Under a U.S.-Russian-Norwegian "Murmansk Initiative," plans exist to increase the facility's processing capacity from 1200 to 5000 cubic meters per year to handle existing and future wastes from decommissioning submarines. [9]

The reduction of this waste would help to prevent future ocean dumping of waste by Russia so that it can adhere to the London Convention's prohibition on such dumping. Discussion of this and other LLRW processing facilities are still under way. Both mobile and stationary treatment facilities are being discussed.

The European Union has funded several studies under the TACIS program. One study examined the feasibility of retrieving spent fuel from the *Lepse* (see Chapter 7). Others are examining ocean flow models of nuclear contamination resulting from ocean dumping.

NATO has also been actively involved in this area through its Committee on the Challenges of Modern Society (CCMS) pilot studies programs. In April 1995, the CCMS worked with military representatives from NATO and former Warsaw Pact countries, including Russia, to develop reports on nuclear and chemical contamination. From these reports, four subtopics have been identified for three-year second-phase studies, three of which directly relate to radioactive contamination:

1. Hazardous constituents in defense-related activities (chemical wastes only; United States lead);

2. Transport of contaminants through rivers, deltas and estuaries (France lead);

3. Safe disposal of radioactive and mixed waste (Norway lead); and

4. Environmental risk assessment for specific defense-related problems (Norway lead). [7]

It is at this point, in June of 1996, that the NATO Advanced Research Workshop (ARW) on "Recycling, Remediation, and Restoration Strategies for Contaminated Civilian and Military Sites in the Arctic Far North" was held. The preceding chapters in this volume were presented at this workshop and served as a forum for further discussion among key representatives from government agencies, industry, and other institutions. Many of the specific recommendations made at the workshop for the disposition of spent nuclear fuel and liquid and solid wastes, as well as for environmental and risk assessments, have been made in other fora. The general process-oriented recommendations regarding project implementation problems that were derived from the workshop are listed below and will be discussed in more detail here because they are problems which can impede the progress of actually carrying out cooperative projects.

1. The ARW noted the importance of developing sound strategies for remediation and restoration of the fragile Arctic environment from contamination by civilian and military activities.

    This is recognized as a problem of international interest. The response should be designed with the understanding that Russia will have the primary responsibility to implement the solutions in the field of nuclear submarine decommissioning and radioactive waste treatment and that these solutions are based on a strategic approach incorporating the contributions of cooperating nations.

    The first task is the identification and prioritization of projects; within these projects, individual tasks for overcoming the problems must also be prioritized.

2. The participants in the workshop recognize the efforts of Russia and Norway to develop a collaborative program on the comprehensive decommissioning of nuclear submarines in northern Europe which began in 1995–1996 and will continue beyond the year 2000 by leading companies in these countries (Kværner in Norway and Energia in Russia, with the involvement of leading Russian firms).

    The intention of the government of Norway to make investments into priority projects under this program is a real step forward in expanding international cooperation to solve the issues of environmental and radiation safety in northern Europe.

    Governments of other countries can also explore the possibility of investment resources for implementing the priority projects related to comprehensive decommissioning of nuclear submarines and management of spent nuclear fuel and radioactive waste generated in the process of decommissioning.

3. It is important that international initiatives and international cooperation are supported. The workshop endorses the ongoing contact and cooperation on this issue between the AMAP/NEFCO study and the IAEA Contact Experts Group (CEG). Specifically, support is given to the present initiative where experts from the AMAP/NEFCO study, together with experts from responsible Russian agencies (MINATOM, Ministry of Defense, Ministry of Defense Industries, etc.), and the CEG Secretariat will discuss and harmonize different proposed projects suggested by the AMAP/ NEFCO study, the CEG, and responsible Russian agencies.

4. The funding of projects is particularly problematic, and it is recommended that a study be conducted of how Russian national funding and international funding can best be coordinated and distributed.

5. Legal, liability, and tax issues are often bottlenecks in reaching accords for international projects. Some effort must be made to resolve the issues of personal and corporate tax exemptions, liability for property and personnel used in the projects, and other legal matters.

6. Each step in a project should be clearly delineated and then monitored and evaluated before proceeding to the next step. This process should be documented to provide a thorough evaluation of each project that will lead to its improvement in the future.

7. Alternative approaches to international cooperative projects should be evaluated according to established cost-effectiveness criteria. These include the integration of existing technologies in order to increase cost-effectiveness.

8. A (formalized) technology transfer program for innovative Russian technologies should be developed. This would include standardized property rights agreements including marketing provisions.

5. Implementation Issues

The first issue discussed was the *need to prioritize specific projects* based on the risks involved and the feasibility of actually completing the projects. (This was done in a general sense in Chapter 1.) Many agencies—both Russian and foreign—had their own agendas and pet projects, and often the larger integrated picture and how each project would contribute to the whole was overlooked. It was recommended that some attempt be made to prioritize these projects—particularly by the Russians, who would have the primary responsibility for all of the projects undertaken.

During the second meeting of the IAEA Contact Experts Group (CEG) for International al Radwaste Projects in the Russian Federation, which was held in Vienna on 10 and 11 September 1996, representatives from MINATOM circulated a draft list of 20 high priority projects. They are listed in Table 1 and reflect many of the projects mentioned before. Some of these projects were initially suggested at the first CEG meeting in May 1995.

It is unclear when these projects were first prioritized and whether there is a general consensus among major Russian actors (including the Navy, Ministry of Defense, and Ministry of Defense Industries) that these are the agreed-upon priorities, but given other studies—especially the Russian and Kværner-Energia reports—and in light of the technical priorities outlined in Chapter 1, this would seem to be a valid prioritized list of proposed projects taking some of the concerns of these other actors into account. The 20 prioritized projects in Table 1 are also consistent with the Norwegian-Russian protocol signed in February 1996.

On 26 September 1996, Defense Ministers from Norway and Russia and the United States Secretary of Defense signed an Arctic Military Environmental Cooperation (AMEC) declaration in Bergen, Norway. AMEC was initiated in March 1995 to discuss Arctic environmental issues that are related to military capabilities and activities. Specific projects under discussion include:

TABLE 1. High Priority Russian Projects (In Priority Order)

Priority Number	Project Name
1	Fabrication and commissioning of special train cars of TK-VG-18 type for transportation of containers with spent fuel from decommissioned nuclear submarines.
2	Construction and commissioning of a storage facility for spent fuel arriving at the Mayak enterprise for reprocessing.
3	Design and construction of facilities for interim storage of spent fuel from decommissioned submarines in the northern region.
4	Design, construction, and commissioning of a vessel for transportation of spent fuel from decommissioned submarines in the northern region.
5	Design, construction, and commissioning of temporary storage facilities for liquid radioactive wastes at the Zvyozdochka plant for utilization [sic] of nuclear submarines in Severodvinsk
6	Design, construction, and commissioning of a mobile facility or facilities for concentration of liquid radioactive wastes from the northern Navy.
7	Design, construction, and commissioning of a temporary storage facility for solid radioactive wastes at the Nerpa plant for utilization [sic] of nuclear submarines (Murmansk region).
8	Creation of a prototype radioactive waste disposal facility at the Bashmachni peninsula at the Novaya Zemlya archipelago.
9	Development of a feasibility study and design on the remediation of the floating technical base-spent fuel storage *Lepse*.
10	Construction of two floating facilities for unloading spent fuel from decommissioned submarines.
11	Construction of industrial liquid radioactive waste treatment facilities in the northern region.
12	Reconstruction of solid radioactive waste storage facilities at the special combine Radon in Murmansk.
13	Creation of a complex of facilities for treatment of solid radioactive wastes from nuclear fleet in Murmansk.
14	Creation of the Northwestern Radwaste Management Center on the basis of the Leningrad Specialized Enterprise Radon.
15	Establishment of a Training-Methodological Center on radioactive waste management.
16	Development of a conception and structure of a normative system regulating safety in radioactive waste and spent fuel management.
17	Development of a sophisticated computer-based system for evaluation of radiation legacy of the former U.S.S.R. (RADLEG).
18	Development of a unified database on radiological situation in the Murmansk region.
19	Development of program and implementation of radiological monitoring of the floodplain of the Yenisei River in the region of influence of the Mining and Chemical enterprise.
20	Radiation characteristics of unloaded spent fuel in the reactor cores of decommissioned submarines.

1. Development of a prototype container for interim storage of spent nuclear fuel.

2. Development of technologies for the treatment of liquid radioactive wastes.

3. Review and implementation of technology for solid radioactive waste volume reduction.

4. Review of technologies and procedures for interim storage of solid radioactive wastes and development of a storage facility.

5. Remediation of hazardous waste sights on military bases.

6. Review and implementation of clean ship technologies.

The first four projects are consistent with the 20 prioritized projects on radioactive waste management that appear in Table 1. The last two cover non-radioactive waste as well as base cleanup methods and the collection and treatment of shipboard wastes and would not be part of the previous activities covering primarily radioactive wastes. Discussions are currently continuing under the AMEC forum concerning the implementation of these projects.

Given some agreement on prioritized projects, the next issue becomes one of *finding enough financial support*—from both Russian sources and cooperating foreign partners, and then assuring that these funds are coordinated and distributed to those parties that can actually carry out the projects. Although many general types of umbrella agreements exist (e.g. the Gore-Chernomyrdin Commission), they are often signed with no financial resources to support concrete activities that might evolve from them. It is important, then, to find pockets of money in various national and international agencies that are specifically earmarked for these activities. Tunold [7] and others have also suggested pooling resources through some kind of international financing mechanism, much like funds for nuclear reactor safety have been pooled through the European Bank for Reconstruction and Development.

After funding is found, the next set of problems arises when private foreign companies begin to work with Russian governmental or quasi-governmental complexes. Contractual obligations, the transfer of hard currencies to Russia's unstable banking system, taxation and liability laws, and protection of personnel are all very difficult issues that arise in the new area of cooperation where little precedent has been established.

Related to the financing issues associated with contractual arrangements are all of the *legal arrangements to protect the rights of the companies* that are using technologies which can be copied by their partners. Mechanisms must be established in which new technologies that may emerge can be equally shared. Incentives must be provided that will enhance the capabilities of all parties involved, instead of the disincentives involved when one side thinks it is being commercially exploited and could lose future profits. Processes and incentives must be developed to produce "win-win" situations out of perceived "win-lose" arrangements.

Finally, as described in the previous sections of this chapter, many countries and international actors have been involved in activities associated with radioactive contamination in northwestern Russia with no prior coordination. *A coordination mechanism* is needed to avoid redundant or unnecessary activities. It has been suggested by representatives of Norway (Tunold and others) at the workshop in Kirkenes and elsewhere [3] that the IAEA Contact Expert Group perform this coordinating function.

Thus, if both funds and activities can be coordinated, it will assure that projects are not replicated, that all actors are aware of each other's activities, and that common standards and practices can be established.

6. Summary and Conclusions

Since the end of the Cold War, an increased emphasis has been placed on assisting the Russian Federation in implementing arms control and environmental agreements and improving the environmental security of the Arctic region through the development of specific remediation projects.

The evolution of this cooperation from environmental assessments of radioactive contamination threats brought about by the decommissioning of nuclear submarines to proposals for concrete projects attests to the slow but steady progress that has been achieved through bilateral initiatives (especially those between Norway and Russia), trilateral arrangements (e.g. AMEC), and international fora (IAEA-CEG). The steady increase in the participation and cooperation of relevant Russian actors is a truly important breakthrough in the thaw in post-Cold Ward diplomacy.

The remediation and prevention of radioactive contamination is a common goal shared by not only Arctic countries, but all countries. The negotiations surrounding Arctic environmental security and the decommissioning of Northern Fleet nuclear submarines can be seen, thus far, as an example of successful negotiations in the post-Cold War period. The goals and projects seem to be clearly defined, at least in this area, and some consensus reached on projects that can improve regional environmental security.

It now becomes the task to provide the coordination and funding necessary—from the Russian Federation first and foremost and from cooperating partners—to complete these projects.

174

References

1. AMAP Expert Group. (December 1995) *Proposals for Environmentally Sound Investment Projects in the Russian Part of the Barents Region, Volume Two: Radioactive Contamination,* The Nordic Environmental Finance Corporation (NEFCO) Barents Region Environmental Program.

2. Bergman, Ronny; Baklanov, Alexander; and Segerstahl, Boris. (May 1996) *Overview of Nuclear Risks on the Kola Peninsula: Summary Report,* International Institute for Applied Systems Analysis, Laxenburg, Austria.

3. Godal, Bjørn Tore. (29 October 1996) *Statement to the Storting on Nuclear Safety Issues,* (unofficial translation).

4. Government Commission on Matters Related to Radioactive Waste Disposal at Sea ("Yablokov Commission") created by Decree No. 613 of the Russian Federation President on 24 October 1992. (1993) *Facts and Problems Related to Radioactive Waste Disposal in Seas Adjacent to the Territory of the Russian Federation,* Small World Publishers, Inc., Moscow, Russia. Translated by Paul Gallagher and Elena Bloomstein.

5. Nilsen, Thomas and Bohmer, Nils. (1994) *Sources of Radioactive Contamination in Murmansk and Arkhangelsk Countries, Volume 1,* The Bellona Foundation, Oslo, Norway.

6. Strand, Per and Cooke, Andrew. (1995) *Environmental Radioactivity in the Arctic,* Scientific Committee of the Environmental Activity in the Arctic, Østerås, Norway.

7. Strand, Per; Salbu, Brit; Christensen, Gordon C.; Lind, Bjorn; Selnæs, Tone D.; Rudjord, Anne Liv; Føyn, Lars; Nikitin, Alexander I.; Chumichev, Vladimir B.; Valetova, Nailia K.; and Shkuro, Valentina N. (November 1995) *Dumping of Radioactive Waste and Investigation of Radioactive Contamination in the Kara Sea: Extended Summary,* Joint Norwegian-Russian Expert Group for Investigation of Radioactive Contamination in the Northern Areas, Østerås, Norway.

8. Tunold, Betzy Ellingsen. (26 June 1996) *Nuclear Activities in Northwest Russia: Selected Projects Within the Norwegian Plan of Action,* presented at the NATO Advanced Research Workshop, "Recycling, Remediation, and Restoration Strategies for Contaminated Civilian and Military Sites in the Arctic Far North," Kirkenes, Norway.

9. United States Office of Technology Assessment. (1995) *Nuclear Wastes in the Arctic,* United States Government Printing Office, Washington, D.C.

10. Wamke, Paul C. and Earle, Ralph II. (1996) SALT II and Beyond, *International Negotiation* 1, 213.

INDEX